Prediction and Perception of Natural Hazards

Advances in Natural and Technological Hazards Research

VOLUME 2

The titles published in this series are listed at the end of this volume.

Prediction and Perception of Natural Hazards

Proceedings Symposium, 22–26 October 1990, Perugia, Italy

Edited by

J. NEMEC
Consultant United Nations,
Department of Humanitarian Affairs,
Geneva, Switzerland

J.M. NIGG
University of Delaware,
Newark, U.S.A.

and

F. SICCARDI
University of Genova,
Genova, Italy

International Decade for Natural Disaster Reduction

1990 - 2000

KLUWER ACADEMIC PUBLISHERS
DORDRECHT / BOSTON / LONDON

Library of Congress Cataloging-in-Publication Data

Prediction and perception of natural hazards / edited by J. Nemec and J.M. Nigg and F. Siccardi.
p. cm. -- (Advances in natural and technological hazards research ; v. 2)
Proceedings of a workshop held in Perugia, Italy, Oct. 22-26, 1990.

1. Natural disasters--Congresses. 2. Natural disaster warning systems--Congresses. I. Němec, Jaromír. II. Nigg, Joanne M. III. Siccardi, Franco. IV. Series.
GB5001.P73 1993
363.3'472--dc20 93-4585

ISBN 978-90-481-4289-7

Published by Kluwer Academic Publishers,
P.O. Box 17, 3300 AA Dordrecht, The Netherlands.

Kluwer Academic Publishers incorporates
the publishing programmes of
D. Reidel, Martinus Nijhoff, Dr W. Junk and MTP Press.

Sold and distributed in the U.S.A. and Canada
by Kluwer Academic Publishers,
101 Philip Drive, Norwell, MA 02061, U.S.A.

In all other countries, sold and distributed
by Kluwer Academic Publishers Group,
P.O. Box 322, 3300 AH Dordrecht, The Netherlands.

Printed on acid-free paper

United Nations

IDNDR
1990 - 2000

THE INTERNATIONAL DECADE FOR NATURAL DISASTER REDUCTION

OBJECTIVES OF THE IDNDR

Reduction of the loss of life, property damage and social and economic disruption caused by natural disasters.

SPECIFIC TARGETS OF THE DECADE

By the year 2000, all countries, as part of their plan to achieve sustainable development, should have in place:

(a) Comprehensive national assessments of risks from natural hazards, with these assessments taken into account in development plans;

(b) Mitigation plans at national and/or local levels, involving long-term prevention and preparedness and community awareness; and

(c) Ready access to global, regional, national and local warning systems and broad dissemination of warnings.

ORGANIZATIONAL ARRANGEMENTS

Special High Level Council (SHLC)

The Council is a body consisting of internationally prominent persons who:

* advise the Secretary General on the Decade;

* take action in promoting public awareness and mobilizing the necessary support from the public and private sectors.

Scientific and Technical Committee (STC)

A group of 25 scientific and technical experts selected on the basis of their personal capacities and qualifications, and with regard to the diversity of disaster mitigation issues and equitable geographical representation.

The functions of the committee are:

* to develop overall programmes to be taken into account in bilateral and multilateral cooperation;
* to assess and evaluate the activities carried out in the Decade and to make recommendations on the overall programmes in an annual report to the Secretary General.

The Committee also includes experts from the various organizations and bodies of the United Nations system.

National Committees

108 national committees and focal points have been established around the world. They form an inter-disciplinary and inter-sectoral group with a key role in the realization of Decade targets.

Their roles include:

* identification of national priorities;
* formulation and implementation of disaster mitigation and prevention plans;
* developing and monitoring of projects;
* providing information and advice to governments, public service and related organizations and to the general public.

A number of regional conferences of national IDNDR committees have taken place to date, and a world conference of national IDNDR committees which will undertake a mid-term review of Decade activities is scheduled to take place in 1994.

Secretariat

Established at the United Nations Office in Geneva, in close association with DHA, the Secretariat is responsible for the coordination of Decade activities and providing Secretariat support to the Special High Level Council and the STC.

TABLE OF CONTENTS

PREFACE TO THE WORKSHOP PROCEEDINGS

JOANNE M. NIGG
Department of Sociology and
Disaster Research Center
University of Delaware
Newark, DE 19716
U.S.A

FRANCO SICCARDI
University of Genova
Hydraulic Institute
Via Montallegro 1
16145 Genova
Italy

The International Decade for Natural Disaster Reduction provides a unique opportunity for scholars and professions from a variety of disciplinary backgrounds to address the major issues confronting national and local governments as they attempt to reduce the vulnerability of their social and built environments from the impacts of natural disaster agents. One of the ways to reduce the impact of a disaster agent is to warn those who are potentially endangered by a particular agent in order for those who are at risk can take protective actions.

In order for warning systems to be effective, they require three basic components:

(1) a detection system to monitor changes in the physical environment that can be identified as constituting an unusual threat to life and property safety;

(2) a system to transmit this information to appropriate officials who are responsible for safeguarding public health and welfare; and

(3) a system to transmit information regarding the heightened threat to the endangered public.

The first component, the detection system, involves the **prediction** of the imminent occurrence of some potentially harmful hazard agent--a flood, hurricane, tornado, or earthquake for example. Physical scientists in a variety of disciplines are working on these detection systems, hoping to increase the amount of time communities have to prepare themselves for the impact of and response to an imminent, dangerous event. The second two components involve the translation and communication of this scientific information into a form that is usable by the intended audiences, whether they are public officials and emergency response organizations or the general public. The manner in which these potential users understand the messages they are receiving is related to the **perceptions** of threat and endangerment that emerge in any warning situation.

Under the sponsorship of the National Science Foundation[1] of the United States and the National Research Council of Italy[2], a meeting of over forty scientists from the fields of geophysical sciences, atmospheric sciences, engineering, and the social sciences was held at the Water Resources Research and Documentation Centre (WARREDOC) of the Italian University for Foreigners in Perugia, Italy from October 22-26, 1990. The IDNDR Secretariat supported the attendance of over 20 additional observers from various developing countries who contributed to the discussions of the formal presentations.[3]

This workshop was unique in that it brought together scholars from several disciplines who had expertise in natural hazard research. While similar efforts have taken place in the past, bringing together European and American scholars, this effort acknowledged the need to bring together physical scientists, social scientists, and engineers who address different, yet interrelated, aspects of the warning process.

The general purpose of this workshop was to address these interrelated issues of prediction and perception as they affect the development of effective warning systems. Two specific objectives were addressed by the scientific presenters: (1) the identification of methodologies for predicting disaster events across different natural hazard agents; and (2) the identification of the factors associated with appropriately interpreting the hazard warning and taking protective action.

Five thematic areas were discussed at the various sessions of the Workshop, including:

1. Large-scale atmospheric phenomenon (hurricanes and tornadoes);

2. Large-scale geophysical phenomenon (earthquakes and volcanic eruptions);

3. Floods and landslides;

4. The response of user communities to scientific information; and

5. Media presentations of scientific information.

In each area, discussion focused on the **uncertainties** in the detection and prediction processes, and how these uncertainties are responded to by decision makers and the general public. An attempt was also made to understand differential response patterns due to different cultural, economic, social, political, and experiential contexts within which scientific predictions are made and warnings issued.

With respect to each of the above topics, the Workshop participants addressed the following three hazard-related issues:

1. The ability of physical and atmospheric scientists to detect and predict potentially disastrous natural hazards;

2. The use of that resultant prediction or forecast information in existing warning systems in different countries; and

3. The variation of response by decision makers and the general public to prediction or forecast information due to variations in perception of threat and the effectiveness of the transmission of the information.

The workshop was organized to maximize interaction of the participants, both within formal sessions and in less formal situations. During the formal sessions, panelists summarized their presentations, leaving the majority of time to an open discussion with the other workshop participants and attendees.

On the final day of the workshop, extensive small-group discussions were held to summarize what had been learned during the workshop that might have application for the IDNDR. Each of the three small groups contained scholars from both the physical sciences and the social sciences, with expertise in different natural hazards.

The scientists, with full participation of the invited observers, determined that it would be useful if the following recommendations, both for research and application, were undertaken during the IDNDR enhance disaster reduction:

1. A current body of knowledge pertaining to disaster preparedness and mitigation already exists; and efforts to adequately transfer this information and respective technologies are more important than increasing the scope of fundamental (or basic research) in the physical and engineering sciences. To the extent that research activities in the physical sciences should be undertaken, they should concentrate on those areas where "uncertainties" still exist in the ability to detect and predict future natural disasters.

2. An international glossary of hazard terms would be useful to facilitate communication among scientists from different disciplines and different countries. If the IDNDR is to make substantial progress toward hazard

reduction, such a glossary would indeed be useful since multidisciplinary efforts will be necessary to address disaster-related problems.

3. In an effort to both transfer knowledge and to enhance compatible data collection among scientists in different countries, it is recommended that a network of existing centers should be established. In regions where no such centers exist, interested scientists and decision makers could "link in" to the network to gather assistance from existing centers of excellence (ie, those centers that are known for their research and application efforts).

4. Regional centers should be encouraged to develop cross-national and multidisciplinary pilot projects to assess hazards, vulnerability, and capacity for both warning and disaster response.

5. Critical assessments of models of risk perception and risk response should be undertaken cross-nationally to understand the decisionmaking process under conditions of uncertainty in order to enhance the development of warning and prediction (or forecast) response systems.

We wish to end this Preface by acknowledging the efforts of Dr. Daniel N. Adom of the Water Resources Research and Documentation Centre (WARREDOC) and the staff of the Prof. Joanne Nigg at the Disaster Research Centre in the preparation of the manuscripts of the proceedings.

FOOTNOTES

---

[1] **The participation of the social scientists who attended this workshop was supported by Grant No. BCS-9016380 from the National Science Foundation.**

[2] **The National Group for the Prevention of Hydrogeological Disasters of the Italian National Research Council supported the participation of a number of Scientists to the workshop.**

[3] **This was the first official scientific activity to take place under the United Nations' newly approved International Decade for Natural Disaster Reduction (IDNDR).**

SESSION 1

THE EFFECT OF THE UNCERTAINTIES IN NATURAL HAZARD PREDICTION ON THE USER COMMUNITIES

A NON-STRUCTURAL POLICY FOR THE MITIGATION OF FLOOD EFFECTS: THE ARNO PROJECT

FRANCO SICCARDI
Hydraulic Institute
University of Genova
Via Montallegro 1
Genova
Italy

DANIEL N. ADOM
Water Resources Research & Documentation Centre
Villa La Colombella
Colombella Alta
06080 Perugia
Italy

INTRODUCTION- A few questions of semantics

" Quando si parte il gioco della zara
colui che perde si riman dolente
repetendo le volte e tristo impara" (1)
[Dante, Div. Comm., Purg. VI.1.]

In the streets of Florence in the thirteenth century, though prohibited by law, dice players used to shout *"zara"* when the total value of the three dices rolling over the table showed the lowest probability sets of results (less than or equal to seven, or greater than or equal to fourteen, meaning *"null try"*): "quia cum tribus taxillis raro veniunt sex vel quinque et rarius quator vel tria...eodem modo accidit inter quator decim et decem octo" (2). From the arabic *"azzahr"*, the dice, the sense of rare events, not involving any special connotation of danger or risk, came into French as *"hasard"*, into Italian as *"azzardo"* and into English as *"hazard"*.

In such a sense the word *"hazard"* is now used by the scientific community, irrespective of the common use, (see, for example, the Longman Dictionary of Contemporary English,

(1) When the dice play is over the looser stands downcast and sad trying again and unhappily learning.
(2) "...because with 3 dices six or five seldom appear and very seldom four or three...likewise it happens between fourteen and eighteen. From the XIV century latin translation and commentary of the Divine Comedy of Fratris Jhoannis De Serravale.

J. Nemec et al. (eds.), Prediction and Perception of Natural Hazards, 3–12.

Longman, 1982, pag. 520) where hazard is thought synonymous of danger and a hazardous event connotes an occurence which involves risk or danger.

It should be stressed that a natural hazard is a rare natural event which occurrence, in an environment created by men used to coping with more frequent events, will most probably cause a disaster due to its character of being a rare event. It is obvious that a hazard, in case it occurs without impacting on human communities, does not contain risk or danger.

A few more remarks about how we could possibly agree on words such as *"prediction"* or *"perception"*, and we shall move out of semantics:

- *prevision*, or *forecast*, is the result of the efforts put into knowing in advance what will happen, in space and time, based on previous experience and/or stringent laws about the ways of real world systems.
- *prediction* is the result of pre-announcing (to someone) a future event, i.e. of transferring (to someone) the forecast;
- finally, *perception* is the result of perceiving, meaning the empirical and/or intellectual knowledge of an exterior real object, and casting and organizing it into a synthetic mental image.

More intriguing questions

Predicting a natural hazard and perceiving the corresponding risk require two types of people:

- *scientists* forecast the harmful event (because a human community is involved) and make up a prediction, i.e. a message containing the relevant information;
- the *endangered human community* receives the message, perceives the risk, i.e. makes up a mental image of the future impact and its costs, and takes actions.

The chain can fail, and frequently fails in developed and developing countries of the real world, due to three possible reasons:

- failure of the scientists to correctly forecast: meaning their inability to locate the harmful event, in space and time, and to measure the uncertainties involved;
- failure of the scientist in predicting, in building up a message, correctly transferring the whole content of information with their uncertainties;
- failure of the human community in perceiving the complexity of the message, which results in the inability to take the suitable actions.

Improving any link in the chain improves the overall reliability. A few years ago scientists in Italy decided to make an effort to face such challenging questions, cross-fertilising from different disciplines the individual disciplinary areas: people from differing physical and engineering science disciplines together with people from the social sciences and a number of local authorities from Italian regions were called into play.

The National Research Group for the Prevention of Hydrogeological Disasters of the Italian National Research council, has established a number of interdisciplinary projects to enhance the applied research activity in the field of disaster mitigation as well as to improve the perception of Italian natural hazards (Rossi & Siccardi, 1991)

CHARACTERISTICS OF DISASTROUS FLOODS IN ITALY

Climatic characteristics

Various analysis of disastrous floods that have occurred in Italy during the last fifty years show the peculiarity of rainfall and basin geomorphology characteristics of the northern-Mediterranean region. The storms producing catastrophic floods have generally extremely high rainfall intensity: the maximum daily precipitations often attains 30%, and sometimes even 40%, of the mean annual maxima. Little is known about the storm-generating mechanism for extreme events, which develop at the synoptic scale with a strong influence of the relief which attains high elevations at short distances from the coast line.

Outlying storms are characterised by low variability of rainfall intensity in time and space, and by steeper growth curves, with increase in return period, compared to storms producing the ordinary annual maxima.

Basin geomorphological characteristics

Flood events present very different features according to the geomorphology of the basins. Excluding the Po, the largest Italian river, with a maximum basin area of 70000 km^2, rivers have upstream basin areas included between 2000 and 16000 km^2. Therefore, the mean response times of the hydrographic network are short: 10-20 hours. In addition, because of high sediment transport, extreme floods cause short-term erosion and sedimentation, and hence instability and changes in channel and flood plain configurations.

The flooding events in the mountain streams have quite a different feature, because they have smaller basin areas, steep slopes, and hence very short response times (order of magnitude of one hour). Unlike plain rivers, the main responsibility of flooding by mountain streams cannot be always ascribed to water discharge above critical values, but often to conspicuous deposits of sediments which occur under particular circumstances in the stream bed. During outlying rainfall storms, because of landslides from hillslope and/or debris and mud flows from steeper tributaries, the sediment supply to main stream is enormously greater than the transport capacity, so that the sediments tend to deposit along the stream, with more striking effects at confluence sites (Di Silvio, 1988). Backwater effects of natural or artificial narrowing, or slope changes, reduce abruptly the sediment transport capacity and form obstructions or natural dams, with the risk of lateral spilling or overtopping with erosion or breaking, as during the Valtellina flood in July 1987. The overaggradation of the bed in many cases does not only cause a rising of water levels, but can even bury buildings and structures located near the stream under a thick mass of mud and debris.

Historical background of flood control problems

Italy has had a long history of flood protection and soil conservation schemes which dates back to the Roman period. Until recently, flood control schemes have been carried out mainly by structural measures which aim at improving the land drainage of alluvial plains, and the hydraulic capacity of major rivers. After the Tiber flood of 1870, the last of the many and disastrous inundations that Rome suffered in its long history, the historical centre was shielded by large side walls and channel improvements. Along the Po river between 1801 and 1876, as many as 214 levee breaks were recorded; due to river training works and levee rehabilitation from 1918 up to date, in seventy years, the breaks have been only 63. As a consequence of structural works the Italian territory became much safer from flood risk. In addition, it is noteworthy to mention that in the last century more than 2 million hectares of marshes were reclaimed.

Nevertheless the disastrous Po flood of 1951 brought to the fore, once again, the issue that whatever the design flood an exeedence risk does exist. The following year, a plan of soil conservation and flood control works, still based only on passive measures, was passed by law. However, in a lecture given in 1952, Giulio De Marchi pointed out that the long systems of embankments cannot represent a final and safe solution to the problem of protection from floods because reinforcements and raising of levees did make disasters less frequent, but not prevent them; when the critical discharge is exceeded and defence works fails abruptly, the inundation can assume the importance and the character of a national disaster.

After the Florence flooding in 1966, a Committee for the study of flood control and soil conservation problems, presided over by De Marchi, was set up to update the 1952 plan of structural works with a broader view, taking into account all relevant technical, economical, legal and administrative problems. At the experimental level a small number of telehydrometric gauges were set up on larger rivers, and, as an integral part of the existing and planned structural works, non-structural measures in the form of emergency procedures and flood warning systems, were considered necessary and beneficial.
However, because of the short hydrological response time and the heavy population and development of the flood plains, the lead time from hydrometric measurements is not enough for the social environment to take efficient precautionary measures. For this reason a real time flood forecasting system had been developed in the Italy only for the Po river, based on hydrometric reporting gauges.

FLOOD RISK MAPPING

Risk perception and development of flood prone areas

The attention given to risk assessment has been increasing during the past decade, with special emphasis on perceived nuclear plant hazards and in the area of toxics and carcinogens. On the other hand in the area of natural risk, mainly floods, society seems to be memoryless. Catastrophes hit unprepared communities. Introducing the concept of risk into the developing processes of urban and industrial areas is a very difficult task, because the fact that an area is prone to be catastrophically flooded once or twice in a century is not perceived as a real economic hazard: restrictions on land use, or re-development actions, are viewed as pure costs without benefits. This fact influences the decisions of the elected local or regional authorities, which tend to be biased toward long term huge structural measures, rather than non structural ones.

The discussion inside the Group during the early meetings to design a research policy recognized that a negative loop connects the lack of political decisions about risk mapping non-structural risk mitigation policies and the risk unconsciousness of the land users: the willingness to pay of the land users is very low, because they do not perceive the economic value of the risk. In this way the gap between funds needed and funds raised for the flood protection continues to grow, and preparedness is compromised.

To try to break the negative loop, the Group made the decision to choose a few national areas where the problem of flooding unprepared communities is great and to use them as field examples for an eventual national policy.

A FULL SCALE EXPERIMENT THE ARNO PROJECT

Aims of the project

The Arno project spans over socio-economic issues such as:

- preparedness of the technical services of the municipality
- to put into operation in real time flood proofing measures;
- improving the consciousness of risk in the population, opening the way to private and public insurance against the risk;

and over technical issues such as:

- improving the reliability of flood peak estimates through ground-based, real time rainfall observations
- predictability of the extreme events based on meteorological radar and satellite observations

Area of interest

The structure of the Arno river valley results in a cluster of primordial lakes, now quite totally filled up by sediments, connected by restrictions of the valley. In the recent past the flood plain where Florence had been developed was covered by the remnants of a former lake leaving the swamp known as "le padule": at the end of 1800's the completion of land reclamation gave to the whole area of Florence the present aspect. The historical records of city flooding start from 1177: in table I a data base of historical floods is provided, where the magnitude levels are evaluated on the basis of produced damages. For purposes of comparison, the two catastrophic floods of the past and present century are included and are shown in brackets. After the 1966 flood, a number of river works had been undertaken, but, due to the excessive amount of funds required for a substantial reduction of the risk, there is a general agreement that, during the next few decades, the Arno river in Florence would be able to convey without overtopping the walls only floods with a return period on the order of 100-150 years. This value has to be considered totally unacceptable if compared with the historical and economic value of the city.

Civil Protection Plan

The components of the Arno project are many faceted. The social issues are grouped into the civil protection plan, which provides for the format of the warning messages to be diffused conditional on the hydrologic alert, and describes the precautionary measures to be taken by those people managing museums, churches and monumental or historical buildings, to put in a safe position paintings, books, and valuable artistic pieces in as short a time interval as possible. It lists also the precautionary measures that needed to be taken by the technical services like aqueducts, gas and electric supply, telephone services, traffic control and so on. Each component of the plan has a prescribed behaviour in the period before the possible flooding and in the period after, if the event does take place. For the purposes of the plan, maps of the inundation prone areas were produced and printed inside the telephone directory. Maps reproduce the areas inundated during the flood of 1966, which is evaluated at a return period of 200 years.

A recent experiment of the preparedness and efficiency of the plan was made during the storm of 24-25 November 1987. After two consecutive days of heavy rainfall, the warning level of the Arno river was reached on the morning of the 24th. A first level warning was issued to alert those authorities responsible for implementing the plan the night before. The second stage of warning was diffused to the population at noon and lasted until midnight. During the night of the 25th the level receeded, and the alert state was cancelled. The Group, with the assistance of the Prefect of Florence, monitored the response of the population, who acted rapidly and efficiently: in less than 6 hours, the actions designed to minimise the flood effects were fulfilled. This experience coupled with the efficient manner in which the risk was perceived greatly encouraged the regional authorities to cooperate in the process of updating the ground based monitoring system, and to give the necessary permissions for the location of a meteorological radar.

DATE	DAMAGE	DATE	DAMAGE
04-11-1177	XXXX	08-11-1550	XX
		13-09-1557	XXXXXX
??-10-1261	XX	31-10-1589	XXXXXX
01-10-1269	XXXX		
15-12-1282	XXXX	??-01-1621	XX
02-04-1284	XXXX	09-11-1641	XX
05-12-1288	XXXX	06-11-1646	XXXX
		??-01-1651	XX
??-??-1303	XX	04-11-1660	XX
??-01-1305	XX	11-05-1674	XX
01-11-1333	XXXXXX	11-10-1676	XXXX
05-12-1334	XXXX	19-02-1677	XXXX
06-11-1345	XXXX	18-05-1680	XXXX
??-11-1362	XX	20-04-1683	XX
01-11-1368	XX	26-01-1687	XXXX
21-07-1378	XX	08-12-1688	XXXX
20-10-1380	XXXX	02-06-1695	XX
		??-01-1698	XX
??-05-1406	XX		
??-12-1434	XX	11-10-1705	XXXX
18-10-1456	XXXX	28-02-1709	XXXX
16-01-1465	XXXX	22-10-1714	XXXX
19-01-1490	XX	06-09-1715	XX
10-06-1491	XX	??-11-1719	XXXX
08-01-1515	XXXX	03-12-1740	XXXXXX
28-08-1520	XX	19-10-1745	XX
15-12-1532	XXXX	01-12-1758	XXXXXX
??-??-1538	XX	15-11-1761	XX
06-11-1543	XXXX		
15-11-1544	XXXX	(12-10-1844	XXXXXX)
13-08-1544	XXXXXX	(04-11-1966	XXXXXX)

Table I: Flood damage in Florence from 1177 to 1761 (Morozzi, 1761)

The ground based monitoring system

The present state of the ground based monitoring system is made up of telemetering raingages and stage recorders upstream Florence: a flood forecasting system based on the routing of the observed flood contributed from the Sieve at Fornacina, from the Arno at Subbiano and from the Arno at Levane allows a lead time in the order of two-three hours.

A simulation performed using the rainfall observations (La Barbera et al., 1988), showed that from ground observations a lead time in the order of four to five hours could be feasible, with a better estimate of the flood peak. The raingauge network is fairly dense and of value for real time calibration of radar measurements. However, as already stated, the performance of the system will no longer exceed a lead time of a few hours, which is less than the time interval required from the social system.

Meteorological radar

The possible use of rainfall images produced by meteorological radar was proposed during the International Conference on Arno Project. A C-band meteorological radar system already available from the Institute of Atmospheric Physics of C.N.R. is operational. The radar coverage was studied (Freni et al., 1987) with special attention to the sub-basins of Casentino, Valdarno Superiore and Sieve, which mainly contribute to flood formation in Florence. The relevance of the orography of the three mentioned sub-basins, comprised of narrow valleys separated by mountain ridges as high as 800 - 1500m, presents limitations concerning the minimum monitorable radar elevation. By the digital elevation model, the best position was first chosen at Monte Maggio (671 m a.s.l.), which is on the left side of the watershed, near Siena. However, due to ecological complaints by "green groups", the final operational position was selected in a second-best location, near Florence.

Digital elevation model

A digital elevation model, of a 400 x 400 meters grid span, is available for the whole Arno basin. The Group provided suitable thematic cartography of the most relevant characteristics of the structure of the terrain to be implemented in such a DEM for hydrological purposes. The available DEM, a large grid scale, was tested recently by the research unit in Florence (Carla' et al., 1987) to check the ability of the DEM to produce a model of the basin useful for hydrological purposes, namely to derive digital drainage and divide lines. Although such a drainage system cannot be used for detailed hydro-geomorphological studies, the partitioning of the whole Arno basin into 800-1200 subbasin units of a minimum size of the order of 1-2 Km^2, is adequate for most applications aimed at building up distributed hydrologic models.

Rainfall models

A cooperative project between the Ralph Parson laboratory of the Massachusetts Institute of Technology and the Civil Engineering Department of the University of Florence was operational to try to answer the major question of the predictability of rainfall by joint use of radar and satellite images with ground observations (Carlà et al., 1989; Ferrara et al., 1991; Filice et al., 1991; La Barbera et al., 1991). The historical records in the Arno basin of the past extreme events, sampled in a large number of raingauges and with very high time resolution has been collected into a data bank (Becchi, 1988) and interpreted with stochastic rainfall models: first results show a good possibility of explaining the morphoclimatic differences between the rainfall processes in the centre of the valley and in the orographic relief (Bras, 1988b; Rodriguez Iturbe et al., 1987). However the predictive ability of such models, stationary in essence, is of no use in real time forecasting, where the prediction must be conditional on the observed storm arrival. Multidimensional non stationary models that describe the storm structure (Bras, 1988a) and update model states through a continuous filtering algorithm in space and time should be necessary. The discussion which followed the presentation of the first results raised a major point about rainfall predictability: if the time scale of a reliable prediction has to be restricted to rain-cell life (Rodriguez Iturbe, 1988), from tens of minutes to an hour, the lead time gain will be of little interest for efficient forecasting; on the contrary, if the time scale of a reliable prediction of the total amount of precipitation over an area reflect the stability of the storm morphology, and, possibly, incorporates into a complete thermofluo-dynamic model the orographic effect, the forecasting horizon will enlarge.

Conclusions

The present review paper emphasizes the research activity carried out by the Italian Group for Prevention of Hydrogeological Disasters. The research themes are coherent with the activity of the international scientific community and a number of ongoing projects will be useful to design cooperative research programs inside the umbrella of the International Decade for Natural Disasters Reduction.

At the national level, a fruitful international cooperation will be of use not only to enhance research, but also to speed up the pace of administrative and technical achievements of the Italian system in the field of soil conservation and water management.

REFERENCES

Becchi, I., Giuli D. (1988), Tecnologie avanzate di misura per il preannuncio delle piene: Progetto Arno, in "Valutazione del rischio idraulico geologico e zonazione; strategie di intervento per la mitigazione degli effetti degli eventi estremi: Rapporto 1987", CNR-GNDCI.

Bras, R. (1989), New approaches to real time flood forecasting: weather radar and fluvial geomorphology in hydrology, in R.L. Bras and Siccardi F. (eds) Selected Papers on "Natural Disasters in European-Mediterranean Countries", NSF-CNR, 415-440.

Bras, R., (1988b), Program report on rainfall analysis issues, M.I.T. cooperative agreement with the University of Florence.

Carlà, R., Carrara A., Detti R., Federici G., Pasqui V., (1987) Geographical information systems in the assessment of flood hazard, The records of the Int. Conf. on the Arno Project, Florence.

Carlà, R., Federici G., La Barbera P., Siccardi F., (1989) Multisensor Analysis of Rainfall Patterns (abstract), AGU Spring Meeting, Baltimora, May 8-12.

Di Silvio, G., (1988) Fenomeni di sovraalluvionamento d'alveo nei bacini alpini, in Previsione e Prevenzione di eventi idrologici estremi e loro controllo - Rapporto 1987, CNR-GNDCI.

Ferrara, A., Filice E., La Barbera P., Minciardi R. Siccardi F., (1991) Precipitation forecasting by multisensor observations, European Geophysical Union, Wiesbaden, April 22-26.

Filice, E., La Barbera P., Siccardi F., (1991) Cloud entities identification from Meteosat images using cluster analysis, European Geophysical Union, Wiesbaden, April 22-26.

Freni, A., Fossi M., Giuli D., (1987) Preliminary results on the analysis of radar siting for Arno Basin monitoring, The records of the Int. Conf. on the Arno Project, Florence.

La Barbera, P., Chander S., Siccardi F., (1988) On lead time of an acceptable forecast at Nave di Rosano, Workshop on Arno River Flood Forecasting, Florence.

La Barbera, P., Minciardi R., Paolucci M., Siccardi F., (1991) Rainfall Pattern Coherence with Multisensors Observations, European Geophysical Union, Wiesbaden, April 22-26.

Morozzi, 1771, as quoted by Becchi, I., (1987) The Arno project: a real time approach to Arno river real time flood forecasting, The records of the Int. Conf. on the Arno Project, Florence.

Rodriguez Iturbe, I., Eagleson D.S., (1987) Mathematical models of rainstorm events in space and time, Water Resour Res., 23 (1).

Rodriguez Iturbe, I., (1988) Chaos and rainfall, Summer School: Hydrology as a Physical Science, University of Genoa.

Rossi, F., Siccardi F., (1991) Coping with Floods: the Research Policy of the Italian Group for Prevention from Hydrogeological Disasters, In: Selected Papers on " Natural Disasters in European-Mediterranean Countries", Bras, R.L., and Siccardi F. (ed) NSF-CNR, 395-413.

THE IMPACT OF STRUCTURAL DAMAGE DUE TO HURRICANES AND THE PROSPECTS FOR DISASTER REDUCTION

A.G. DAVENPORT*
The Boundary Layer Wind Tunnel Laboratory
The University of Western Ontario
London
Canada

INTRODUCTION

There are many factors that can influence the seriousness of a disaster, ranging from the degree of preparedness and planning to the availability of relief and post-disaster assistance. In the case of tropical storms (hurricanes, typhoons and cyclones), however, a primary agency of the disaster is the failure and unservicability of buildings and structures. Amongst these are homes, refuges, hospitals, schools, industrial buildings, communication towers, and power lines. Apart from the personal injuries which the structural damage causes. it deprives people of shelter, disrupts essential post-disaster services such as hospitals and communications, cripples important contributors to the economy, such as the manufacturing, tourist and agricultural industries and impedes relief and recovery. The evidence suggests that most of this structural damage is preventable at little or no cost.

These remarks are illustrated by reference to the impact of Hurricane Gilbert on the island of Jamaica. This storm earned the distinction of being the "hurricane of the century" (an accolade transferred to Hugo a year later). It devastated Jamaica and Cancun in the Yucatan peninsula before finally dissipating in southern Texas. The loss of life was only 45 (11 shot for looting by the police) and injury was low, thanks in part to the effectiveness of warning systems; but the property damage and the contingent losses were severe. In Jamaica the economic losses were roughly 2 to 3 billion ($U.S), equal or greater than the annual GDP.

This paper examines the performance during Gilbert of various types of structure and the impact their failure had on the severity of the disaster. Since improved wind resistance is the key to reducing the disaster, instances are cited where the structures have performed well and their survival factors are indicated. We also examine the protection that codes

* **Chairman: The Joint Committee of the Royal Society of Canada/Canadian Academy of Engineering on the International Decade for Natural Disaster Reduction**

J. Nemec et al. (eds.), Prediction and Perception of Natural Hazards, 13–21.

provide against storms the intensity of Gilbert. We affirm the need for governmental enforcement of codes in the design of structures essential to post-disaster service; but the implementation of codes more widely is likely to require, in addition. the support of other sectors and organizations. We draw attention to the special needs of the construction industry for the skills and information to improve disaster resistance.

The storm

"Tropical Storm Gilbert" became "Hurricane Gilbert" on the morning of Sunday, September 11th 1988 when it was still six hundred miles from the eastern extremity of Jamaica. It travelled towards Jamaica at a steady speed of approximately fifteen miles per hour and reached the island at approximately 11am on September 12th. The central pressure was 963mb, and the eye diameter (roughly twice the radius to maximum winds) was about 25 - 30 miles.

The storm track lay along the centreline of the island, as can be seen in Fig. 1, and the region of strong winds affected all parts of Jamaica. The eye left Negril at the westerly extreme at approximately 8am on September 13th.

Windspeeds in the storm were monitored by aircraft from the NOAA National Hurricane Centre, Miami as well as anemometer recordings in Jamaica at Norman Manley Airport, Kingston and Sangster Airport, Montego Bay. Although the recordings were interrupted by damage to the instruments, the evidence suggests that in open areas close to the coast the sustained windspeeds (averaged over about 10 minutes) at 10 metres height were approximately 85 to 95 mph while the peak gusts were in the range 130 to 140 mph. Windspeeds on the north coast, to the right of the track, were somewhat higher than on the south coast.

These windspeeds are similar to the design windspeeds in Jamaica; that specified by the Jamaican Building Code is a 3-second gust speed of 125mph), and that specified by CUBIC, the Caribbean Uniform Building Code is a "once-in-50-year" 10 minute mean wind speed of 83mph; higher Windspeeds with longer return periods are appropriate for structures strategically important in post-disaster recovery. It seems Gilbert very closely represented the "design windstorm".

Terrain and topography can significantly modify the windspeed as is suggested in Fig. 2. Winds near the surface approaching the land off the sea would first be retarded by the rougher terrain covered by trees and houses. Speed-up would occur over hillcrests and in the lee of hills, there would be a sheltering effect.

Structural damage and its impact: an overview

The following summarizes the nature and extent of some of the structural damage during Gilbert and the impact it had on the recovery after the storm:

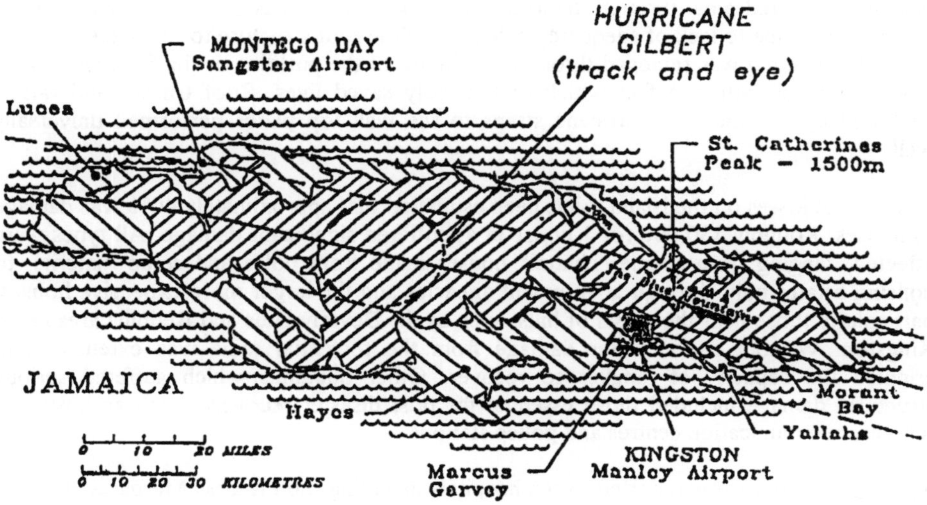

Fig.1: Track of Hurricane Gilbert in Jamaica

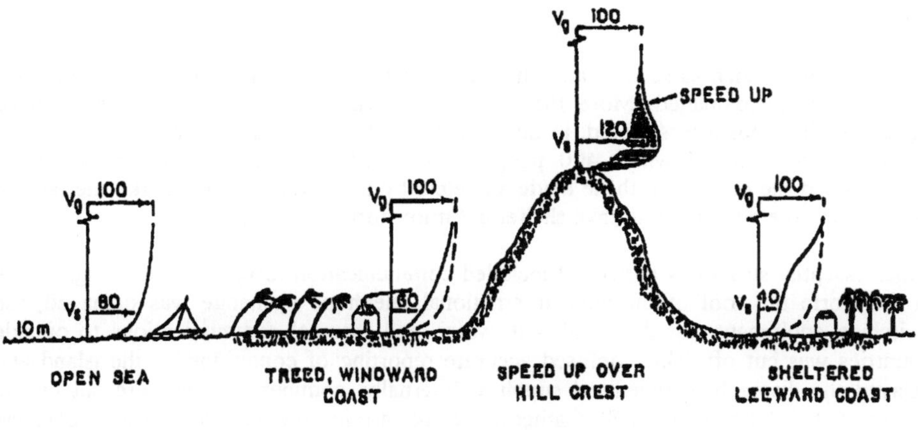

Fig.2: Influence of Terrain and Topography

Buildings suffered most through the loss of roofing. In some cases the entire roof system was removed due to lack of adequate anchorage of the roof structure to the walls. In other cases the sheeting was removed through the failure of the purloins or the fasteners. Walls stood up much better, a fact which undoubtedly saved lives. Roof trusses and rafters anchored by the standard hurricane straps to the wall top plates performed universally well.

Of the total losses on the island it has been estimated that approximately 40% was damage to houses. Roughly 130,000 (or 1 house in 4) suffered significant damage. All types were affected, ranging from simple "chattel" houses made from poles and scrap material, to housing estates built through government agencies and the larger more expensive houses, particularly those near the crests of hills, such as the Beverly Hills and Redhills areas near Kingston. Without roofs, water damage from the torrential rains was extensive and crippled the capacity of families to recover. Other buildings which suffered serious structural damage included those essential for post disaster recovery, such as hospitals, refuges, communication centres and police stations.

Significant damage was reported to ten hospitals including the Princess Margaret Hospital at Morant Bay, a hospital for chest diseases. The University Hospital, the new hospital at Montego Bay and the hospital at Mandeville. In each instance, the community was faced not only with the replacement of the structure, supplies and costly medical equipment, but also with the loss of facilities to treat those injured in the storm and the patients already admitted for care.

Other buildings, such as schools and churches which were designed and used as refuges, were also badly damaged. More than 500 of the islands 580 schools were reported damaged. Churches also suffered; in one instance in Kingston, an entire roof was blown off a church inside of which 400 people were sheltering. These experiences further compounded the distress of those made homeless by the storm and will make the task of encouraging the population to take shelter in future storms, more difficult.

Other essential structures destroyed included communication towers and buildings. Early in the storm the roof of the main international telephone exchange was damaged, the switching equipment was drenched with water, and external communications to outside countries was cut off. This confused accurate reporting of conditions on the island and delayed the dispatch of relief and supplies. Internally, communications were cut by the failure of the 300ft. tower at St. Catherine's peak carrying the main microwave repeaters for the island. Towers at the police headquarters in Kingston and at the military base at Newcastle were destroyed, interfering with the essential military and police communications. Other towers at radio and television stations on the island were also damaged, preventing broadcast of warnings and bulletins. As is often the case, amateur radio operators filled the communication gap during a critical period.

The attractive Mona campus of the University of West Indies is situated in a funnel shaped valley above Kingston which seemingly intensified the winds, particularly those from the north during the initial phase of the storm. Roofs were removed from the Administration Building with the loss of the University records, the Performing Arts Centre, the principal student residence, and the Law faculty. The losses included the Law and other library collections (some unique national resources), valuable research equipment and results, the loss of teaching space and a delay in commencing the academic year.

Public utilities such as power and water were interrupted for many days - weeks in the inland regions. Although the main high voltage distribution network on the island. which was carried on steel towers. survived the storm intact. 50% of the wooden utility poles were destroyed both by wind, and fallen trees and branches. Steel light standards were extensively damaged. Water supply was interrupted in many regions, in one instance, for example, due to the collapse of a roof over a reservoir facility.

There was extensive damage to industrial buildings throughout the island. Principally these were older buildings but there were numerous examples of newer buildings which had been destroyed. The loss of these structures had a direct impact on the productivity of the economy and jeopardised both the income of the workers and the acquisition of capital to re-establish the industry.

Another major industry affected was the tourist industry, the island's largest foreign currency earner. Numerous hotels, particularly those on the north coast from Ocho Rios to Montego Bay, were seriously damaged. Fortunately, the storm surge was negligible which might have taken heavy toll of beach hotels, and the damage that took place was almost entirely due to wind and rain. Photographs which appeared in the foreign press shortly afterwards of hotels without their roofs caused tour operators and individual vacationers to switch their bookings.

Although damage to larger office buildings in downtown Kingston was relatively light, there was extensive glass breakage. Because of this, water damage to computer systems and records was in several instances serious. One insurance company lost records on its policyholders.

Losses to the agricultural industry contributed significantly to the measure of the disaster. In addition to the very heavy crop damage - to bananas, citrus, sugar, coffee and coconut palms - there was widespread damage to structures such as storage sheds, and chicken houses (the occupants of which were decimated). This left a community which was accustomed to being self reliant for food, suddenly dependent on imports.

Devastating as the effects of Gilbert were on Jamaica, nonetheless, a number of ameliorating factors lessened its impact. Unlike hurricane Charlie, the storm surge accompanying the storm was slight and there was no serious flooding: however, some warnings of flooding were issued based on incomplete information and turned out to be false alarms (a fact which could jeopardise evacuation measures in future hurricanes).

Reference has been made to the excellent advanced weather warnings although-communication of these through press and broadcasting may not have been fully effective in all instances, there was nevertheless sufficient time for most of the population to take precautions and lay in essential supplies. This again saved lives and reduced losses.

A major factor in the recovery was the relatively high level of insured risk - approaching 40%. These funds, unlike some funds earmarked for reconstruction, had no strings attached, were available quickly and could be used for a wider range of needs. A further advantage, was that early 99% of the risk was reinsured in hard currency, a fact which allowed for the early purchase of essential construction materials from hard currency sources. There has been some suggestions that the high failure rate may make it more difficult in the future to obtain reinsurance on the same favourable terms.

A further important factor in the recovery was the generous aid from the surrounding Caribbean community, by Jamaican communities in Canada, the United States and the United Kingdom as well as the national governments of these and other countries. The logistical tasks of distributing this aid was formidable and again hampered by the failure of structures at the airport, (including the radar facilities), and communications in the island.

The Prospects For Reducing The Disaster

The evidence suggests that the root cause of this disaster was the failure and unservicability of structures due to wind action. Improved wind resistance is therefore the logical starting point in any effort at disaster reduction.

Amongst the essential requirements to achieve this goal is the establishment of appropriate construction codes or standards, education and training in their use, and their application in practice. In many instances, such as in buildings critical for post-disaster services and recovery, the enforcement of design standards in new construction and the upgrading of old construction is essential. This is not easy and may require special efforts to achieve. Close inspection of the damage in Gilbert showed that the achievement of adequate wind resistance in many instances would have required only minor design modifications, negligible costs in materials, but better inspection.

With respect to construction codes, Jamaica is fortunate in having available not only the Jamaican Building Code (first published in 1983), but also the recently introduced Caribbean Uniform Building Code (CUBIC). Both of these are suitable codes, however, neither has been officially enacted and their effectiveness is considerably reduced.

The CUBIC code is directly based on the new ISO standard for wind resistance. It incorporates a recent study on the risk of hurricane windspeeds in the Caribbean, approaches to determining the modification to the wind speeds due to terrain and topography, the influence of dynamic response of the structures to wind turbulence, and recent wind tunnel studies carried out in state-of-the-art "boundary layer" wind tunnels. The provisions for the dynamic response and the speed up over hills would have both been highly significant in the definition of the appropriate design windspeed for the St. Catherine's Peak tower.

Reconstruction of a number of the failure mechanisms of failed structures suggested that nowhere did the wind pressures required to cause failure exceed those specified in the CUBICŠ code. This was important reassurance of the adequacy of this code.

Standards for roofing materials exist in the literature, but, judging by the wide range of materials and techniques employed and the widespread losses, do not appear to have reached all contractors and influenced practice sufficiently. The development of suitable standards and construction guides may require further attention. At least one very effective set of guidelines for the construction of small houses has been prepared.

Aluminium sheeting suffered the most extensive damage. The advantage of this material is that it is mined and manufactured locally and its cost is highly competitive with steel sheeting and other roofing systems. Reasons for the losses include the use of grades of material which were brittle and prone to tearing at the heads of fasteners; the gauge of the material was too thin; fasteners lacked washers, were too few or were inappropriately spaced; purloins were spaced too for apart; and there was insufficient attention to pressures at eaves and overhangs. These deficiencies can be addressed straightforwardly but to do so requires the framework of effective standards, and where these are not available, appropriate testing. Some of the responsibility for providing these standards must rest with the industry itself.

There were many examples where good wind resistant practices were put into effect. High voltage transmission lines was one. Another was the widespread practice of reinforcing block walls (introduced largely as an earthquake counter-measure but reconfirmed by experience in Hurricane Charlie) which was again shown to be effective and undoubtedly saved lives. The use of hurricane straps to hold down rafters to wall top plates was simple and effective. (If omitted, the radio slogan warned "strapless today, topless tomorrow").

Other roofing practices performed well. This includes the use of concrete slab roofs - a style introduced after Hurricane Charlie in 1951 - as well as other traditional roofing practices - some using tongue and groove planking, some with steeper pitches and "hipped" construction at the ends. Contemporary methods often omit the planking and use laths in their place, and the roof pitches are flatter leading to higher uplift forces. Recent wind tunnel tests have confirmed the superiority of the hip roof over the gable roof, both in terms of its more benign distribution of wind pressures and its greater structural efficiency in redistributing the pressures.

An outstanding illustration of hurricane resistant design concerned a group of about twenty identical pre-engineered metal buildings (PEMB's), built to CUBIC wind resistant standards. These buildings were situated at "industrial free zones" in different surroundings throughout the island (see Fig. 1 for localities). All were subjected to winds near the maximum intensity or sites which were in open exposures to the wind. Although in each case there was extensive damage in the surrounding regions, none of the PEMB's were damaged, and their use was uninterrupted They were an example of an international assistance project built to an appropriate standard of wind resistance. in Gilbert and in the past, many such projects were not.

Conclusions

A key to the reduction of disaster due to hurricanes like Gilbert is the improvement in the wind resistance of structures. Steps towards this goal include the establishment of standards, education in their use and application in practice. In Jamaica several excellent standards have been prepared but their translation to practice requires further work. Additional work may be required for example in professional equation and in developing simple manuals and instructing in their use.

To implement these standards will require the co-operation of a number of participants and the development of a strategy that has the support of the community. Enforcement of standards through legal enactment may be appropriate for structures essential for post-disaster services andŠrecovery; the application to other types of structure will require the additional support of other sectors and agencies, including the government planning and housing agencies, the manufacturing, tourist and agricultural industries, the insurance industry, the banking and lending agencies, and finally the development banks and bilateral aid programmes. Each of these organizations has a vested interest in better construction standards and through their concerted efforts, persuasion, and closer technical monitoring of their own investments, will collectively be able to achieve these standards.

Finally the construction industry itself, including the architectural and engineering professions, the contractors and the manufacturers has special needs to ensure that what is designed and built can withstand hurricane winds. This vital industry is fragmented and often neglected; it has with shortages of skills and training for new technologies, of equipment and "user friendly" technical information. Addressing some of its needs is an important aspect of disaster reduction. One of the important purposes of the international Decade for Natural Disaster Reduction, is to foster this collective action and make prevention more effective.

Acknowledgements

These remarks are based on visits to Jamaica following hurricane Gilbert. The writer acknowledges assistance received in preparing this paper from many individuals, including Messrs. Franklin MacDonald, Garth Kiddoe, Alfrico Adams, Wayne Reid, Tony Gibbs, Al Wason, David Allen, Peter Georgiou, and Mike Mikitiuk, as well as attendees at a two day seminar jointly sponsored by the Jamaican Institution of Engineers and the Canadian Society of Civil Engineers.

References

Allen, D.E., (1989). "Report on damage to structures in Hurricane Gilbert", Institute for Research in Construction, National Research Council.

Davenport A.G., P.N.Georgiou, D.Surry (1984). "A hurricane wind risk study for the Caribbeun with special consideration of the influence of topography," Boundary Layer Wind Tunnel Laboratory report to the Pan Caribbean disaster Preparedness and Prevention Programme, United Nations Disaster Relief Organization, 1984.

Davenport, A.G., (1989). Proposed new international wind load standard, The Keynote speech, Second Asia-Pacific Symposium on wind-engineering, Beijing, China.

Mecchanm, D., D. Surry, A. G. Davenport, Wind action on hip and gable roofs, Colloquium on Industrial Aerodynamics, Aachen

Proceedings of a Seminar on the Effects of Hurricane Gilbert on Buildings and Engineering Facilities in Jamaica; Jamaican Institution of Engineers and The Canadian Society of Civil Engineering, Kingston Jamaica, 16-17 February, 1989.

THE SOCIAL DIMENSIONS OF WARNING AND THE TRANSITION FROM FOLK WISDOM TO LAYMANSHIP

WOLF R. DOMBROWSKY
Institut fur Soziologie
Christian Albrechts Universitat
Olshausenstrasse 40/60
D-2300 Kiel 1
Germany

When reflecting on appropriate IDNDR-strategies, we must seek relief programs that avoid dependency and inactivity. To put it positively, we must seek programs that support indigenous efforts to prevent disasters and to overcome hazards without continuous outside support.

In practice, better relief strategies are frequently limited by several characteristics. These include their ad-hoc nature, narrow efficacy, and short comings in integration with other programs (e.g. economic aid, IWF). Other characteristics include incompatibility with national policies/doctrines and international controls (terms of trade, protectionism, GATT) and the lack of empirical, theoretical, and political certitude. Specific uncertainties are based upon the differences between so-called "developed" and "less/least developed" countries (LDCs/LLDCs), upon the high percentages of the international aid budgets that flow into the economies of the "givers" for producing and shipping aid-appliances and for maintaining their organizational and professional frameworks, and upon the many international (disaster) relief programs that more often undermine the moral and independence of the afflicted populations instead of fitting for self-help initiatives and an appropriate development on the part of the "receivers."

Until today we lacked conclusive models of the interrelations between natural and human factors. Most theories on large-scale, long-term change are still contested. Thus, most relief-projects are contradictory in diagnosis and therapy and, as a consequence, mutually prevent improvement. Often, the complexity and interdependency of problems, the plurality of scientific analyses, and the withdrawal from a distinct causality toward theories of probabilities will be utilized for exculpating perpetual misconduct.

Many uncertainties are particularly based in our concepts of modernization and industrialization. In the view of all the advantages and benefits, the disadvantages and risks are often overlooked, sometimes ignored. The external costs of industrialization especially

J. Nemec et al. (eds.), Prediction and Perception of Natural Hazards, 23–28.

are not taken into account. Nevertheless, the idea that underdevelopment will be surmounted by exporting the model of free market economy, Western technology, and private business is familiar. The idea disasters will be avoidable by implementing sophisticated safety measurements, by modernizing the modes of production, transportation, and communication, and by intensifying education and training is familiar as well.

Case studies, retrospective analyses and other evaluation studies of economic aid and of past disasters prove these ideas to be myths. Despite all well-intentioned efforts, the gap between rich and poor has widened; the number and strength of disasters have increased all over the world. However, the type, seize and effects of disasters seem to correspond with the distribution of wealth: two thirds of all people who will be killed annually by geophysical hazards are living in the poorest parts of the world (with a per capita income below 1000 US$/year), whereas the material losses are preponderantly in the wealthier parts. Obviously, the reproduction of "modern" societies depends on the use of technology and commodity values whereas the reproduction of "pre-modern" societies depends upon the use of physical strength (i.e. human labor force, animals and natural energy sources) and agricultural products.

The conclusion, however, that fewer people would die if the poorest countries of the world industrialize is absolutely wrong (see Clausen/Dombrowsky 1987). Industrialization was and is accompanied by other, specific hazards that annually produce millions a victim as well. The distinction between "natural" and "man-made" hazards is a misconception because the use (or misuse) of nature is literally "man-made", regardless of the standard of civilization. Thus, it is not an earthquake that destroys a city or a drought that destroys the harvest; it is a failure in land-use, settlement, building construction, agriculture, infrastructure, knowledge, and so on (see Dombrowsky 1989).

Historically, "hazards" have accrued from (grievous) experiences. The threats of nature and enemies required protection and defense. Errors and failures demanded remedy and durable betterment. Generally speaking, remedial measures and betterments are survived experiences and surmounted failures that are incorporated into the practical knowledge, into the conception of the world (animistic/magic/scientific), and into the material culture (new techniques, solid tools). Consequently, the standards of civilization go hand in hand with an appropriate texture of knowledge, competencies, and capabilities.

Traditional knowledge about threats and dangers is based on the close contact with animate and inanimate nature. In contrast to "modern man", the "pre-modern man" is able to read and interpret the "signs" of fauna and flora, weather and waters. All over the world a huge body of folk wisdom can be found that hands down a variety of narratives on the "behavior of nature" in the face of danger (especially before earthquakes), on extraordinary changes or irregularities in plant behavior, in atmospheric lights and colors, in movements in air and waters, and on animal behavior (see Geenen 1990). This body of knowledge built the traditional "disaster culture", appropriate behavior pattern, reasonable precautions and deducible adequate warnings.

With industrialization, the accord between disaster culture and vulnerability is lost; the knowledge of dangers and threats is no longer appropriate. But in contrast to industrial societies, no direct or indirect substitutions are available. The advanced instruments of direct disaster protection correspond with the technological standards of production, transportation and communication in use. Both are beyond the understanding of the people who are living outside the industrial and commercial areas of LDCs/LLDs. The indirect safety and insurance systems of industrialized societies - protective labor legislation, environmental protection, social insurance, public health care and the system of public education - are only partly found or not found at all in LDCs and LLDCs.

In societies where entire families or clans are supported by a few members, (economical) rationalization is inevitably leading to pauperization and poverty. Without functional equivalents to modern social insurance systems, people are forced to undersell their labor force, switch to semi-legal or illegal activities and exploit their females and children. Consequently, in pre-modern societies the substitution of labor force by modern technology is leading to an "immiserizing growth" (Griffin 1974) on the societal level and a de-functionalization and depreciation on the individual labor-force level.

Moreover, industrialization without a mitigating, sociable embedding in public education, vocational training, and general adjustments to science & technology will also lead to the de-functionalizing and depreciating of traditional skills, knowledge, and kinship. In contrast to modern societies, in pre-modern societies knowledge originates predominantly from learning by doing and from imitating what is done by the older people. However, the skills and capabilities required by the process of industrialization are not at hand and, consequently, cannot be transmitted to the growing generation. On an objective level, the traditional knowledge, the social structures of age and relationship, and the modes of socialization become inappropriate for mastering the new demands of an industrial society.

Without adequate competencies and organizational structures, most families try to maintain their traditional settings and routines, although they lose their original functions. Urban settlement and the slums around industrial areas (like Bophal) demonstrate that the ties of kinship and traditional living are more durable than the exigencies of markets, mobility and flexibility. On the individual level, however, the dilemma still remains and becomes a vicious circle. The traditional family is no longer capable of preparing the growing generation for getting a job, but without earned income, the survival of the traditional family is endangered.

Focused on the support of their families, most people are concerned with the direct, daily survival. In the cities and the industrial areas, their traditional knowledge is almost worthless. Without adequate education only a job as casual- or day-laborer is available. Mostly, people have no insight into the significance of their job or in its risks. In contrast to the knowledge of their rural life and its threats, people are blind to urban and industrial dangers Ä and, consequently, part of the danger. Conclusively, the industrialization of LDCs/LLDCs may not only cause an "immiserizing growth" but also the miserizing of unsafety. The idea that "safety" foremost depends upon the immanent conditions of technology is a typical misperception of engineers and technicians. More failures, accidents, and disasters are prevented with handy, smart, improvised, circumspect, prospective, clever, and socially responsible interventions of "normal people" than with instructions, automatic controls or redundant systems.

From this point of view, developed and underdeveloped countries are looking into the same abyss. The social competencies to avoid damage, harm and failures are an important part of the "human capital". Modernization in terms of progressive division of labor is leading toward a radical specialization and, consequently, toward the dependency from specialists. Similar to the de-functionalizing and depreciating of traditional competencies in LDCs/LLDs, the human capital in industrialized societies is going to be devaluated by professionalism. In each special field everybody knows more than ever before, but in an emergency, when general capabilities and social capacities are required, the lack of knowledge, know-how, and adequate information will aggravate the peril.

Another misconception is the definition of "disaster" as an event especially concentrated in time and space. Epidemiologists, for example, know that disastrous plagues may be under way but view them as chronic diseases. A serious outbreak will be called a disaster, although the illness has been present for some time. This is obviously true for AIDS, for desertification, and for nuclear melt downs. All these events are unexpected, because their precursors are seldom recognized as warnings.

In science-based societies, many attempts are made to identify precursors rationally. Sophisticated surveillance and control systems (satellites, remote sensoring, TA) have been implemented to foresee dangerous situations. In this respect, every warning system is an endeavor to be faster so that the future will become the present. Thus, time is the most important variable when people try to avoid failures.

The severity of a disaster is related to the speed of preparation for the disaster. If the danger is faster than any protection, wreckage is unavoidable; if readiness is as fast as the onset of danger, withstanding is possible; and if readiness is faster than the onset of danger, even precautionary measures can be utilized or safer grounds can be reached. Hence, disaster may be defined as result of interfering periods of time, or, in other words, as proportion of correlating speeds. The speed which people need to analyze their situation is the first factor that will influence the rapidity of an upcoming danger. Stunned people "accelerate" the onset of danger; fast reactions "decelerate" the onset of danger. Consequently, threats have no speed of their own, no absolute, unevadable velocity. Thus, it is the human reaction to risk and danger that makes the choice from the spectrum between total failure and total safety.

The loss of sovereignty of time is equivalent to pressure: Without the time to weigh one's chances, deciding is at random and therefore far below optimum. The loss of sovereignty of space is equivalent to narrowness. Without the chance to maintain distance from danger and threat, fear takes command and makes people feel literally trapped and driven into a corner. In terms of decision-making theory, this is known as "deciding under stress and incomplete information". Although (tolerably) incomplete information is normal in every-day life, under stress the time to react and, consequently, the speed to reach other places is additionally diminished. Conversely, sufficient information may enlarge time and space because intellectual sovereignty procures internal (emotional, psychological) distance and instantaneous readiness. Thus, information is a "functional equivalent" of time and space.

The speed of a warning depends on the cultural standard of alertness; and the significance of signals depends on the risk-perception and the risk-taking behavior of all people. Therefore, warning-signals need a specific receptivity, an auditory that is instantly and autonomously capable of interpreting the contextual meaning of a warning. Consequently, the most important aspect of warning is not the warning itself but the societal standard of general alertness. Only the acquaintance with the risks and threats of one's civilization, a competent risk-perception and risk-assessment and an appropriate stockpile of knowledge and protection-measures led to a permanent alertness and an operative preparedness. Both are preconditions to realizing warning-signals and to surviving dangers. More important than that is the attempt to maintain the positive potential of the "human capital" to avoid hazards. Without this potential, even the specialized professionals are useless. Thus, it is a question of survival to adjust the quality of the "human capital." Without a widespread folk wisdom embedded in a proven disaster culture, the risks of modernization and industrialization are severe.

REFERENCES

Journal articles

Castleman, B. I. (1979) "The Export of Hazardous Factories to Developing Nations." Journal of Health Services 9 (4):569-606.

Clausen, L. and Dombrowsky, W. R. (1987) Katastrophen, in: Pipers Wörterbuch zur Politik, Bd. 6 "Dritte Welt. Gesellschaft, Kultur, Entwicklung," ed. by D. Nohlen and P. Waldmann. München, Zürich: Piper, 264-270.

Book references

Burton, I., Kates, R. W. and White, G. F. (1978) The Environment as Hazard. New York: Oxford University Press.

Dacy, D. C., Kunreuther, H. (1969) The Economics of Natural Disasters: Implications for Federal Policy. New York: Free Press.

Dombrowsky, W. R. (1989) Katastrophe and Katastrophenschutz. Eine soziologische Analyse. Wiesbaden: DUV.

Geenen, E. (1990) Risiken and Chancen von Erdbebenvorhersagen aus katastrophensoziologischer Sicht. Untersuchungen zu ihren Bedingungen in einem Risikogebiet in der Westturkei. DFG-Forschungsbericht Kiel 28/8-1. Kiel.

Green, S. (1977) International Disaster Relief. Toward A Responsive System. New York: Mc Graw-Hill.

Weede, E. (1990) "Rent-Seeking or Dependency as Explanation of Why Poor People Stay Poor." Globilization, Knowledge and Society. Albrow, M. and King, E. London, Nebury Park, New Delhi: Sage/ISA 173-192.

White, G. (1974) Natural Hazards: Local, National, Global. (ed.) New York: Oxford University Press.

Technical reports

United Nations Disaster Relief Organization: Disaster Prevention and Mitigation. A Compendium of Current Knowledge, UNDRO, 11 Vols., New York:

Vol. 5, 1978: Land Use Aspects
Vol. 7, 1979 (a): Economic Aspects
Vol. 10, 1979 (b): Public Information Aspects
Vol. 11, 1984: Preparedness Aspects

NATURAL HAZARD PREDICTIONS AND RESPONSES IN VERY LARGE CITIES

JAMES K. MITCHELL
Department of Geography
Rutgers University
New Brunswick, NJ 08903
U.S.A.

Introduction

This paper focuses on natural hazard prediction and associated responses in the context of large scale modern urbanization. It reviews evidence from a selection of large-city earthquake, storm and wildfire disasters and focuses on implications of the Loma Prieta earthquake of October 17, 1989. It is suggested that, in large-city disasters of the future, issues of prediction and response are likely to be over-shadowed by other pressing management concerns.

Urbanization and disasters: a brief overview

Urbanization is one of the most important factors propelling worldwide growth in natural disaster potential (Cuny 1983; Havlick 1986; National Research Council 1987). People and material investments are pouring into cities that are already exposed to significant physical risks, or are expanding into areas at risk, or are pushing against the limits of biogrophysical systems and sociotechnical systems. Among in-migrants there is a general loss of indigenous hazard-copying mechanisms that were better suited to rural settings but may be inappropriate in the expanding cities (Ward 1990).

In recognition of this situation, assessment of the "vulnerability of megacities" has been identified as a high priority task for scientific investigation during the International Decade for Natural Disaster Reduction (IDNDR) 1990-2000 (International Ad Hoc Group of Experts 1989). It is of particular interest to Japanese experts who were among the Decade's first proponents and are its most steadfast supporters (Earthquake Engineering Research Institute 1985; Committee for Disaster Research of the Science Council of Japan 1989; National Land Agency 1989; United Nations Center for Regional Development 1990).

J. Nemec et al. (eds.), Prediction and Perception of Natural Hazards, 29–37.

For the purposes of this analysis, the term "very large cities" is synonymous with "million cities" (i.e., cities with populations in excess of one million people). It includes "megacities" (e.g., Mexico City, Sao Paulo, Tokyo), as well as other of lesser size.

Previous research

The scale, speed and complexity of urbanization are neglected topics in most investigations of hazard prediction and response (Drabek 1986; Sorensen, Vogt and Mileti 1987; Vogt and Sorensen 1987). The few studies that have examined these subjects focus on the experience of small medium-sized cities (Clarke, Curson, Kayastha and Nag 1989; Kosinski and Zelinsky 1991). There is a lack of historical evidence about disasters in very large cities, mainly because such cities are a recent phenomena.

In 1800 London became the world's first million city. By 1900 there were 13 million cities. Now the numbers have grown to between 170 and 300 with most of the increase occurring in the past fifty years (UN 1987). The world's largest city (Mexico City) currently contains at least 10 million people. The entire contiguous urbanized area of Mexico City (i.e., the central city plus its adjacent smaller cities, towns and suburbs) may hold close to 20 million. It is projected that there will be at least two urbanized areas of 20 million by the year 2000 (Mexico City, Sao Paulo) and another 20 of between 10 and 20 million (Dogan and Kasarda 1988).

Recent disasters in very large cities

Before World War II, the Tokyo-Yokohama metropolitan area was the only very large city that had been devastated by a natural disaster. The great Kanto earthquake (1923) killed 105,000 people and destroyed 702,000 houses (Mizutani and Nakano 1989). At that time the metropolitan area contained around 2,765,000 inhabitants. Between 1945 and 1988, at least 92 such disasters have been reported in data published by the U. S. Office of Foreign Disaster Assistance (Figure 1). This represents between 1 percent and 2 percent of all recorded disasters during that period.

In addition to Tokyo/Yokohama, four very large cities are known to have sustained major loss of life and heavy material damage since World War II: Nagoya, Japan (1959); Tangshan, China (1976); Bucharest, Romania (1977); and Mexico City, Mexico (1985). Many other large cities have experienced lesser--but nonetheless substantial--losses. They include places like Honolulu (1982), Houston/Galveston (1983), Adelaide and Melbourne (1983), London (1987) and the San Francisco-Oakland area (1989; 1991). Five of these cities were affected by earthquakes, four by severe storms and three by wildfires.

Evidence from past disasters - earthquake disasters

Earthquakes have caused the heaviest loss of life and the greatest devastation in very large cities. Earthquakes that struck Tokyo/Yokohama, Tangshan, Bucharest, Mexico City, and

Loma Prieta were either not predicted or not well predicted and no warnings were issued. As a result, people did not seek shelter in earthquake-resistant buildings and no evacuations occurred in advance of these events. However, a recent study indicates that warnings of aftershocks that were issued in the wake of the Loma Prieta earthquake, were successful in triggering a wide range of protective behavior among a majority of local residents (Mileti and O'Brien 1991).

So far as is known, neither the Kanto (Tokyo/Yokohama) earthquake, the Bucharest earthquake, nor the Mexico City earthquake, were predicted or forecast. Chinese experts claim to have predicted the Tangshan earthquake one month before the event, but failed to follow up with detailed warnings in the days before the disaster (United Nations Center for Regional Development 1990, p. 84). In July 1980, USGS scientists estimated that there was a 50% probability of a major earthquake in the San Francisco Bay area within 30 years and highlighted a seismic gap in the Santa Cruz mountains. When a moderate earthquake occurred there in mid-1988, it attracted considerable local media attention. Another moderate earthquake took place in the same vicinity during August, 1989. Geologists issued an advisory that a larger quake might follow within five days, but non occurred. Some observers believe that heightened preparedness in the wake of these predictions, combined with experience gained from a large emergency exercise conducted ten weeks before the earthquake, paid off when the Loma Prieta earthquake of October 17 struck (Schultz 1990; Baumgardner 1990).

Severe storm disasters

Severe storms have produced a spate of major emergencies in very large cities that involved great disruption as well as heavy material and economic losses (e.g., Nagoya; Honolulu; Houston/ Galveston; London 1987). The Ise Bay typhoon (Nagoya 1959) was a watershed event in Japanese disaster history. Its death toll of 5,000 has not been surpassed by any subsequent disaster and the losses it inflicted triggered a major reappraisal of Japanese disaster policy. In 1982, Honolulu and the metropolitan island of Oahu were affected by the most costly natural disaster to strike Hawaii during recorded history (Mitchell 1985). Hurricane Alicia (National Research Council 1984) was the third most costly storm to strike the United States. In 1987, a severe windstorm triggered London's most serious public emergency since World War II (Mitchell, Devine and Jagger 1989).

In Nagoya, Honolulu and Houston/Galveston, the oncoming storms were tacked--and their landfalls predicted--at least 24 hours before they came ashore. In contrast, the storm that affected London was not detected until a few hours before it arrived. Few members of the British public received effective warnings and fewer still took precautions. Where warnings were issued, limited evacuations took place and they probably contributed to reducing loss of life. Nonetheless, the level of evacuation from urban areas was low and emergency preparedness systems were often flawed. For example, preparedness measures were not implemented in the Ise Bay (Nagoya) region because the typhoon struck after

6:00 p.m. on a Saturday--a time when residents were at home and many professionals were not on duty (United Nations Center for Regional Development 1990). A hurricane warning was never issued for Honolulu and residents continued with normal activities throughout the storm. In Galveston, though warnings were issued, the city was not evacuated, possibly because of concern over a previous evacuation that was perceived as having been "unnecessary."

Wildfire disasters

Catastrophic wildfires have burned into suburban neighborhoods of several very large cities in Australia including Adelaide, Melbourne and Sydney (Bryant 1991). The Ash Wednesday 1983 fires that affected Adelaide and Melbourne are widely regarded as the worst natural disaster to have affected Australia. fire conditions were forecast several days before they developed and emergency managers were on alert. The scale of the subsequent conflagration exceeded all expectations and heavy losses ensued. According to one commentator "No modern technology, no prescribed burning, no preparation by homeowners or municipal councils, no improvement in prediction, no numbers of firefighters or amount of equipment could have stopped those fires from occurring or negated their effects" (Bryant 1991, p. 168).

Urban fringe wildfires are becoming an increasing problem in California and elsewhere in the United States (Pyne 1991). Two dozen deaths and property losses in excess of $1 billion were sustained in Oakland, California after the city's eastern edges were burned on October 20, 1991. Although fire conditions were forecast and the initial blaze was brought under control, it flared up again before firefighters could be redeployed. Subsequently the fire invaded city neighborhoods at high speed, compelling many people to flee immediately.

Assessment

Given the limited information that is available from these case studies, it is difficult to assess the role of prediction, warning, evacuation and predisaster sheltering in very large cities. It is clear that predictions either were not made, or were not followed up, or were followed up with inappropriate responses. These cases may be examples of missing or failed responses to hazards. It is not possible to determine how much loss might have been prevented by improved prediction and response systems, or how many other communities have been adequately protected by similar adjustments.

Though many lives have been saved by the use of warning and evacuation systems in the United States and similar results are being achieved in rural regions of less developed countries, it is not clear that the same benefits will easily accrue in large cities of the Third World. It is the sense of this observer that even when accurate and reliable short-term predictions and responses are obtainable (and, in the case of earthquakes, they are probably still some way off), they will not address the most serious disaster problems that arise in large Third World cities. In almost all of the study sites, the worst loss of life and the heaviest economic losses, were due to: the collapse and burning of structures, the impairment of transportation facilities, and the severing of lifelines. By themselves, prediction and response systems may not be as successful in big Third World cities as elsewhere because other essential parts of the hazard-mitigation system are missing. For example, wholesale evacuations of Third World city populations are probably not feasible because it is difficult to find suitable shelter in fringing shantytowns or rural hinterlands.

For partial evacuation and sheltering schemes to be effective, it will be necessary to target specific communities and vulnerable groups in high risk areas. Detailed local microzonation maps of risks will be required together with directions for emergency preparedness measures and the use of appropriate nearby shelters. Few urban communities in the Third World now possess these capabilities. In addition, few buildings in Third World cities are constructed to the hazard-resistance standards of buildings in developed countries. Hence, there are fewer reliable shelters available to house populations at risk, if and when sound predictions are made. Nor is it clear how the inhabitants of shantytowns and other poor neighborhoods would be expected to respond when presented with information about impending risks. For example, should they leave their homes and move to safe locations in designated buildings or other places, perhaps at the risk of having their houses taken over by squatters or town down by land owners whose property they occupy?

Moreover, the transportation and infrastructure networks that are needed to move and supply evacuated and sheltered populations are less well developed and more vulnerable than similar networks in developed countries. They are intensively used, have fewer redundant components, and are less resilient. A vivid example of these limitations is provided by the city of Guahati in northeast India (Arya 1990). This fast-growing community of 1.5 million people is located near the eastern end of the so-called "chicken's neck," a narrow corridor of land that links Assam with the rest of India. All traffic between the northeast and the remainder of the country passes through Guhati by means of the only two bridges (road and rail) for 165 miles across the wide and unruly Brahmaputra River. Ground transportation for a region of more than 20 million people would be cut off if the bridges collapse during one of the frequent earthquakes or floods.

In short, upgraded prediction and warning systems in Third World cities probably will not work well on their own. To be properly effective, they will require ancillary facilities, personnel and support systems; changes in standards of living and human expectations; linkage to other types of adjustments in local, regional and national hazard protection systems; and the knowledge, resources and willingness to put new procedures and measures into practice.

Many of these findings and suggestions apply to small cities and medium cities in developing countries as well as large cities. But what of the special problems of very large cities? Might we expect new hazard management problems to emerge in the giant cities of the future? The experience of Loma Prieta throws new light on this question because it is the largest disaster to affect a major metropolitan area in the United States during recent years and by far the most costly ever.

Emerging problems in very large cities

The Loma Prieta earthquake affected five million people in the fourth largest metropolitan region of the United States, including the cities of San Francisco and Oakland and scattered smaller communities in and around the Santa Cruz mountains. Four aspects of the Loma Prieta experience are particularly noteworthy: mass media coverage, emergency assistance, infrastructure, and post-disaster recovery. They have been fully reported elsewhere, so only a brief summary of findings is presented here (Mitchell 1990; Webber 1990).

The Loma Prieta experience suggests that: (1) big city disasters are likely to be better reported than their rural counterparts; (2) big city populations may pose new problems for the delivery of emergency response services and disaster relief because they contain complex societal mixes, including minority groups and practitioners of alternative lifestyles; (3) big-city infrastructure networks, such as transportation routes and lifelines, are especially vulnerable to disruption; and (4) the recovery of big cities from disasters is likely to be slower than the recovery of rural areas and smaller communities from similar events, because of difficult-to-solve conflicts about land use and redevelopment.

Hazard predictions and associated human responses are not explicitly mentioned in the above listing because--apart from helping to increase preparedness against aftershocks (Mileti and O'Brien 1991)--they had a relatively minor effect on human responses to the Loma Prieta earthquake. This does not mean that such adjustments might not be important elsewhere or that they are likely to be problem-free in very large cities. However, it might suggest that urbanization parameters are likely to be more relevant to human responses in the emergency period and in the recovery and mitigation phases of disasters, than in the preparedness phase.

Conclusion

The success or failure of hazard prediction and response systems does not appear to have been a major factor in large city disasters of the past. Such systems can offer limited protection to large Third World cities that are at risk to major earthquakes and severe storms, but they require careful coordination with other hazard-reduction measures.

Emerging problems of very large cities in the United States suggest that improvements in relief, recovery and mitigation systems might be more pressing needs for very large cities. This is brought out by recent earthquake and wildfire experience in the San Francisco Bay area where urban development is spreading into naturally risky locations in patterns that are reminiscent of existing Third World cities like Rio de Janeiro or Manila.

Finally, as poor countries become more urbanized, the nature and distribution of disaster losses is subject to change. One possible outcome is that the gap in opportunities for survival that already separates urban and rural communities throughout the world, may become deeper and wider. It should be one task of the International Decade for Natural Disaster Reduction to ensure that this does not occur.

REFERENCES

Journal articles

Havlick, S. W. (1986) "Third Wordl Cities at Risk: Building for Calamity," Environment, Vol. 28, November 6-11; 41-45.

Mitchell, J. K. (1985) "Post-Disaster Prospects for Improved Hurricane Protection on Oceanic Islands: Hawaii After Hurricane Iwa." Disasters, Vol. 9, 286-294.

Mitchell, J. K., Devine, N. and Jagger, K. (1989) "A Contextual Model of Hazard." Geographical Review, Vol. 79, 391-409.

Pyne, S. J. (1991) "The New American Fire." New York Times, October 28.

Book references

Bryant, E. A. (1991) Natural Hazards. Cambridge: Cambridge University Press, 167-170.

Clarke, J. I., Curson, P., Kayastha, S. L. and Nag, P. (1989) Editors, Population and Disaster, Oxford: Basil Blackwell, 24-33.

Cuny, F. C. (1984) Disasters and Development, New York: Oxford University Press.

Dogan, M., and Kasarda, J. D. (1988) Editors, "How Giant Cities Will Multiply and Grow," in The Metropolitan Era. Volume 1: A World of Giant Cities, Mattei Dogan and John D. Kasarda, Eds., Newbury Park: Sage Publications, 12-29.

Drabek, T. E. (1986) Human Sustem Responses to Disaster: An Inventory of Findings. New York: Springer Verlag.

Kosinski, L. and Zelinski, W. (1991) Emergency Evacuations of Cities, London: Unwin Hyman.

Mileti, D. S. and O'Brien, P. W. (1991) Public Response to the Loma Prieta Earthquake Emergency and Aftershock Warnings. Hazards Assessment Laboratory and Department of Sociology, Colorado State University, Ft. Collins, CO.

Mizutani, T. and Nakano, T. (1989) "The Impact of Natural Disasters on the Population of Japan." Population and Disaster, edited by John I. Clarke, Peter Curson, S. L. Kayastha and Prithvish Nag, Oxford: Basil Blackwell, 24-33.

United Nations Center for Regional Development (1990) "Research Project on Urban Disaster Prevention and Management in Metropolit an Contexts" and "Urban Disaster Management Exercise: METRO-DIMEX," in Challenges of the IDNDR: International Decade for Natural Disaster Reduction, Nagoya, Japan, 52-57.

Unpublished papers

Arya, A. S. (1990) "Natural Disasters and Countermeasures in India," presentation to the International Decade for Natural Disaster Reduction: International Conference 1990 Japan, Yokohama and Kagoshima, September 27-October 3.

Baumgardner, E. V. (1990) "The 1989 Loma Prieta Earthquake: Organizational Responses in the State Office of Emergency Services," presentation to the International Decade for Natural Disaster Reduction: International Conference 1990 Japan, Yokohama and Kagoshima, September 27-October 3.

Mitchell, J. S. (1990) "Post-Disaster Living Conditions in Large Cities: Implications of the Loma Prieta Earthquake." Paper presented at the Fourth International Research and Training Seminar on Regional Development Planning for Disaster Prevention, Nagoya, October 4-5.

Schultz, P. (1990) "The 1989 Loma Prieta Earthquake: Immediate Response and Implications for Future Preparedness Efforts." Presentation to the International Decade for Natural Disaster Reduction: International Conference 1990 Japan, Yokohama and Kagoshima, September 27-October 3.

Ward, B. (1990) Comments delivered at the International Decade for Natural Disaster Reduction: International Conference 1990 Japan, Yokohama and Kagoshima, September 27-October 2.

Webber, M. M. (1990) "Inferences for Infrastructure Planning: Effects of the Loma Prieta Earthquake on Transportation." Paper presented at the Fourth International Research and Training Seminar on Regional Development Planning for Disaster Prevention, Nagoya, October 4-5.

Technical reports

Committee for Disaster Research of the Science Council of Japan (1989) Towards a Less Hazardous World in the 21st Century: International Decade for Natural Disaster Reduction - Proposals by Japanese Scientists, Tokyo, p. 13.

Cuny, F. C. (1986) "The Lessons of the September 19-20, 1985, Mexico City Earthquake." Planning For Crisis Relief: Towards Comprehensive Resource Management and Planning for Natural Disaster Prevention, Nagoya: United Nations Center for Regional Development, 157-178.

Earthquake Engineering Research Institute (1985) Proceedings: U.S.-Japan Workshop on Urban Earthquake Hazards Reduction, Publication No. 85-03, Berkeley.

International Ad Hoc Group of Experts (1989) "Implementing the International Decade for Natural Disaster Reduction," report prepared for the Secretary-General of the United Nations, Washington, DC, p. 49.

National Land Agency (1989) Disaster Countermeasures in Japan. Tokyo: Government of Japan, Prime Minister's Office.

National Research Council (1984) Hurricane Alicia, Galveston and Houston, Texas, August 17-18, 1983. Washington, DC: Committee on Natural Disasters.

National Reseach Council (1987) Confronting Natural Disasters: An International Decade for Natural Hazard Reduction. Washington, DC: U.S. National Academy of Sciences, U.S. National Academy of Engineering, p. 10.

Sorensen, J. H., Vogt, B. M. and Mileti, D. S. (1987) Evacuation: An Assessment of Planning and Research. ORNL-6376, Oak Ridge National Laboratory: U.S. Department of Energy.

Vogt, B. M., and Sorensen, J. H. (1987) Evacuation in Emergencies: An Annotated Guide to Research. ORNL/TM-10277. Oak Ridge National Laboratory: U.S. Department of Energy.

THE EFFECT OF THE UNCERTAINTIES IN NATURAL HAZARD PREDICTION ON THE USER COMMUNITIES

RENZO MORCHIO
Physics Department
University of Genova
Via Dodecanneso 33
Genova
Italy

INTRODUCTION

Generally scientists communicate the results of their researches to restricted environments of specialists, well aware of the specialized scientific languages and informed of the limits of the uncertainties implied by the results of such research works. However in some cases scientists must transmit the conclusions of their researches to larger groups of recipients, including people not familiar with the language of science. In such cases, given the fact that the research data utilised are affected by errors and hence are characterised by certain probability values, non-trivial problems of communication could arise.

A typical case of such a situation is often tied to scientific predictions of some natural hazards. In this context, it is a duty of the scientist to inform the competent authorities of the interested area in order that the necessary preventive measures could be assumed. In such circumstances there are problems to translate the scientific language in terms of the current language, but essentially there are problems connected with the fact that the prediction of the catastrophic event is never absolutely certain, with probability equal to 1, but is always given with a given probability value, say p_e. That is not trivial since very often people ask to the scientists for data and predictions that are absolutely certain. It is a myth, of course, but it is a myth that is difficult to eliminate, also in the case of cultured people. A friend of mine, requested by a manufacturer to verify some measures, told me some time ago, that he sent the correct data, each one within an error margin. He was asked what the little numbers preceded by + and - were. He said: the errors. The answer: we pay to have exact results, not errors. For these reasons it seems to be the case to analyze more deeply the situation. The difficulties are connected with the fact that we have to do not only with the uncertainties of the scientific predictions, but also with the uncertainties on how the public authorities involved will interpret the scientific data and what consequently they will do, and eventually with the uncertainties in the reactions of the public opinion. We have to deal with the informative chain reported in fig. 1.

J. Nemec et al. (eds.), Prediction and Perception of Natural Hazards, 39–47.

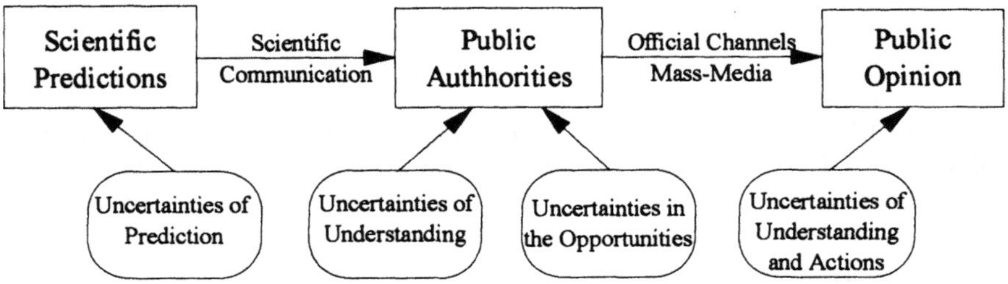

Fig. 1: The rectangles indicate the sources emitting and/or receiving information and the rounded rectangles indicate the sources of uncertainties.

As concerns the first source of information (the scientific predictions): let p_e be the probability of the predicted disastrous event, obviously, $0 < p_e < 1$. Suppose that the event could be localized in a point *0*. Around *0* the effects of the event will manifest themselves with decreasing gravity, with increasing distance from *0*. In this case, at least on a theoretical basis, it is possible to introduce a factor *q*, which indicates, for each point, the predicted gravity of the effects of the event. It seems reasonable to put $0 < q_i < 1$, q_i being the considered factor, written for the point *i*. q_i clearly has the properties of a probability. It is evident that we have spoken and we will speak in the following sentences, of single points to simplify the arguments. In reality we have to do with small areas. We could say that we speak of infinitesimal areas *dxdy*. In this case the q_i have the characteristics of a density of probability.

Around *0*, the points in the plane, each one with its associated q_i, will be a kind of a risk map. Practically the points with very small q_i will be of no interest at all, since for these values the damages will be probably irrelevant. It is as if we would define a small threshold value q_s, below which there are no risks. Considering then only the q_i values for which

$$q_i > q_s \tag{1}$$

we will deal with a circumscribed area. This area, referring to Venn diagrams, could be interpreted as a set. For each point of the plane it is possible in fact, to decide univocally whether or not it is an element of the set, on the basis of the condition in (1).

The q_i distribution in the area described by the set can be different in different cases. For our discussion it is sufficient to adopt one of the possible distributions: probably the most general one. We refer to the gaussian distribution. Considering, to simplify the argument, a section of the considered area crossing *0*, we will have, along such a section, a distribution of the q_i as indicated in fig. 2.

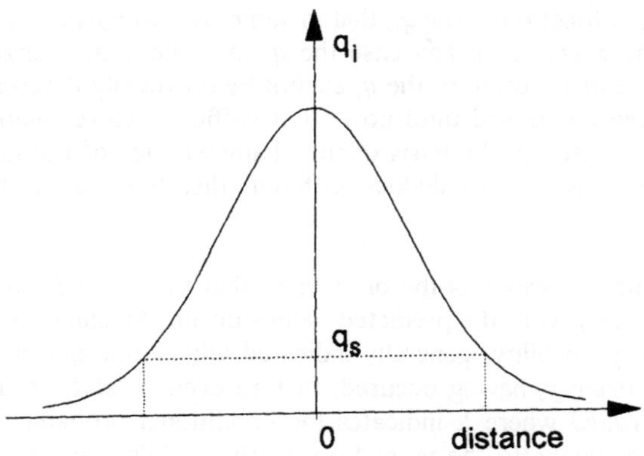

Fig. 2

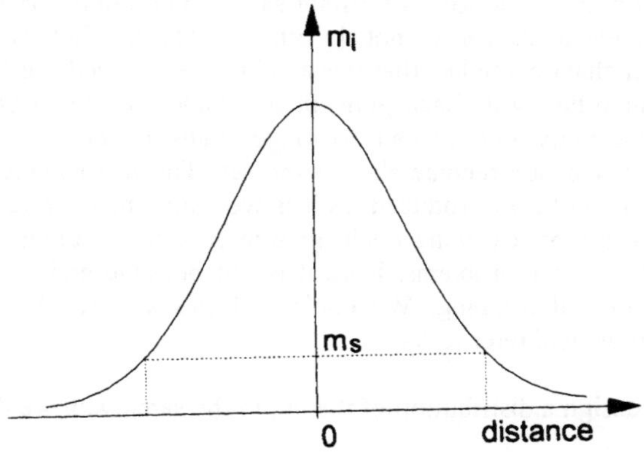

Fig. 3

However the distribution of the q_i described in fig. 2 could be a satisfactory one if the predicted catastrophic event, call it *E*, would be certain, with probability $p_e = 1$. But this is not true. As we know, the event *E* occurs in fact with probability $p_e < 1$. This fact reflects also on q_i. The gravity of the consequences of the event *E* on the point *i* is correctly described by q_i if the event *E* occurs. However, if it is not so, then the q_i also lose their meaning. It seems to us therefore more realistic to work not with the q_i described before, but with a factor m_i, a function of the q_i, that in some way weighs the q_i on the basis of the probability p_e of the event E. In any case the q_i must be characterized by probabilistic components since the distribution of the q_i cannot be univocally determined, although this problem has not been mentioned until now. It is sufficient to remember, as an example, that in any particular case, on the basis of the characteristics of the ground and of other factors, different scientists could deduce different distributions of the q_i that are not coincident.

The m_i could then be expressed as the probability that the event E having occurred with probability p_e, also the q_i with the predicted values occur. As can be seen, we have to do with the conditional probability, generally indicated with a notation as $p(x_i|y_i)$. It means: the probability that event y_i having occured, that an event x_i also occurs. In our case we can write: $m_i = R\ (q_i|E)$ where R indicates the conditional probability[1]. Henceforth we will refer more opportunely, to the m_i and not to the q_i. We can ask, however, what are the consequences of the substitution of the m_i with the q_i. The first relevant consequence is that now the membership of any point in the plane around 0 to the considered set (let's call it S_1), is no more univocally determined. The m_i are tied to probability factors that can change. If in the same way we have done with the q_i, we fix for the m_i too a threshold value m_s, sufficiently small so that all the $m_i < m_s$ do not create any interest and the corresponding points in the plane could be ignored, the m_i verifying the condition $m_i > m_s$ define again a set of points. But now we cannot say with certainty for any particular point whether it is an element of the set, or not. In other words: the fact that the m_i values are probabilistic and can change, implies that these values can exceed the threshold value m_s, but they can also slide below it. Each point in the plane then belongs to the set with a given degree of probability and we can no longer think in terms of a rigid alternative between membership and non membership to the set. The mathematicians know already such situations. For them they introduced, as it is well known, the fuzzy sets (Mandani & Gaines, 1981). This line of reasoning could give interesting arguments for a deeper and analytical treatment of these problems. Here it is not possible and we limit ourselves to underline their theoretical meaning. We could call this second (fuzzy) set S_2. In the following discussion we will refer to S_2.

We can, of course, design a distribution of the m_i, in the same way we did for the q_i in fig.

---

[1] Obviously the reality has been simplified. Generally we cannot have an event restricted in a geometrical point (in 0), although with generalized effects. We will have an area with fuzzy borderline (hatched in the figure) in which the event, in the prediction, cannot be exactly localized. Practically it will determine an area, object of particular attention (the closed area E, in figure). But making the prediction of damages for men, in the case of the catastrophic event, an area of interest Q, will be defined . It will be chosen as a function of characteristics of land, of density of inhabitants, of opinions of the different scientists, and so on. The area Q, in general, does not coincide with the area E and often contains limited places not included in the area E (but included in the fuzzy area). Obviously to the intersection I of the areas E and Q will be attributed the greater attention. The parameters q_i, utilized in this paper, will be defined for each point of I. The fact that I is the area we deal with, justifies the utilization of the conditional probability. It is, in our case: $R\ (qi|E) = R\ (q_i \wedge E)/R(E)$, where R indicates the probability and $R(E) = p_e$.

2. With the same rules utilized for the fig. 2, we can draw the fig. 3. Now each value of the m_i can have a spectrum of values as a function of the different evaluations. This puts us in the best conditions. Suppose that the dispersion of the different values of the m_i is very small and that, taking for each point the average of such values we have again a curve with a gaussian behaviour. In such a case the weight tied to the fact that the event *E* will occur with a probability $p_e < 1$ will play a fundamental role. The effect essentially will be to soften the curve, reducing the maximum value (in *0*). Now the curve will be more sensitive to variations that are also not very strong, determined by other factors: by the influences played by the next sources of fig. 1, as we will see. Since the conditional probabilities are "probabilities", it will again be $0 < m_i < 1$.

The scientific predictions, or in other words, the information coming from the first source of information in fig. 1, in principle, are reduced to:

1. the event *E* is predicted with probability p_e;
2. there is an area interested in the consequences of the event *E*. Such an area could be expressed as a fuzzy set of points whose degree of risk is described by a (probabilistic) factor m_i.

Such information will be communicated to the responsible authorities of the interested area. The behaviour of such authorities will be determined by their understanding of the information received and by the interpretations they will give to such information. There is the problem of not only understanding the scientific language, but also of the interpretation of the values of probability involved. They can give infact in different people not acquainted with the theory of probability different expectations. There is then a problem of interpretation. The derived uncertainty, indicated in fig. 1 by the second rounded rectangle (from the left), could be interpreted and treated as if it were a noise. If the total information transmitted by the scientists were *H(X)* where *H*, as is well known, is the information tied to the transmitted data *X* (the data indicated above, with the number 1 and 2), and $H(X|Y)$ were the information lost for the noise, or the equivocation, to use the term introduced by Shannon (*Y* are the data the responsible authorities believe to have received, when the data *X* have been emitted), the information really transmitted would be $H(X) - H(X|Y)$. It is not easy to evaluate such an expression. But from that could derive a strong distortion of the set S_2. In the best case it could derive from a further attenuation in the distribution of the m_i values. Just to simplify, we can consider a probably very frequent situation: the presence of a "white noise", a noise, to say in simple words, altering in the same way all the transmitted data. In such a case, with reference to the distribution of the m_i as in fig. 3 the consequence could be only an attenuation of the curve (decreasing its maximum), or an accentuation (increasing its maximum). It means that the interpretation of responsible authorities would imply an excessive or too moderate evaluation of the real risks. It could have serious consequences, without any distortion of the information received.

There is however an additional aspect that could be more important of what we have just seen. It is indicated by the third rounded rectangle (from the left) in fig. 1. It is mentioned as "uncertainty in the opportunities". It refers to the fact that in the evaluation of the risk for each point the authorities tend to overestimate political, sociological and economic situations, less important from a scientific point of view. If, as an example, in S_2, in a peripheral side at a small risk, there were an industrial plant suspected to be dangerous and already politically discussed, the authorities, supposed in favour of that plant, could not permit any catastrophic event in that place. Other elements could play a role. A place at high risk, but with a very small density of inhabitants will have for the authorities smaller interest than a place at smaller risk, but with a very high density of inhabitants. If we have to do with elected authorities, we must consider the fact that they will try to estimate the possible reactions of the electors as a consequence of not realized alarms. Such examples could be multiplied. But the results will be always the same. For valid or arbitrary reasons, the responsible authorities will give an high importance to some m_i and a smaller importance to other ones.

In principle, we can formalise this situation too. We can associate to each point of S_2 a factor r_i measuring the interest of the authorities for that point. r_i could be used as a multiplicative factor of the m_i value for the same point. As a consequence for a big r_i the product will be big, but for a small r_i the product will be a smaller one. We will deal with a new parameter m_i', given by:

$$m_i' = m_i r_i \tag{2}$$

To evaluate analytically the consequences of this new step, we would have to know the distribution of the r_i. In general this is impossible. In any particular case, knowing the position of the inhabited places, of the industries and of other things, we could try to do a simulation. But this simulation could not consider the arbitrary factors which are in the head of the politicians. The use of a random number generator, properly programmed, could try to do a simulation also of these aleatory factors, with results probably more interesting from a theoretical point of view, than from an applicative one.

"A priori", as it has been said, the distribution of the r_i cannot be foreseen. It is also difficult to say if the distribution in general is completely at random or it shows some regularities. Probably it will have some (absolute or relative) maxima, with reference to particular places of the considered area. The distribution, then, will present some regularities.

It is not likely, however, that the r_i have the same value in every point and that they work only as attenuants or magnifiers of the curve describing the m_i distribution. In any case, also in the highest regularity hypothesis, the r_i distributions will be bimodal or more likely multimodal. Since the probability that a place largely interesting on a political and social point of view coincides with the point *0*, is small, in general, the consequences of what has been said will be to alter and shift the m_i distribution, presented in fig. 3.

We could suppose that, in a particular situation, the distribution of the r_i is that indicated in fig. 4.

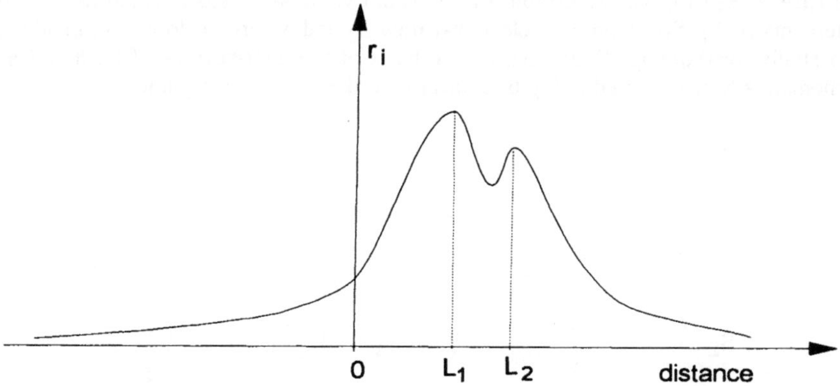

Fig. 4

The points L_1 and L_2 could indicate two places important from a social and political point of view. Such a distribution is quite different with regard to that indicated in fig. 3. If now we describe the situation utilizing the new factors m_i', referring to their definition in (2), we will obtain a new (fuzzy) set, which can be called S_3. It is schematised in figure 5, obtained from the product $m_i r_i$. It is on the basis of this distribution (and not of that in fig. 3) that the responsible authorities, explicitly or implicitly, will take their decisions.

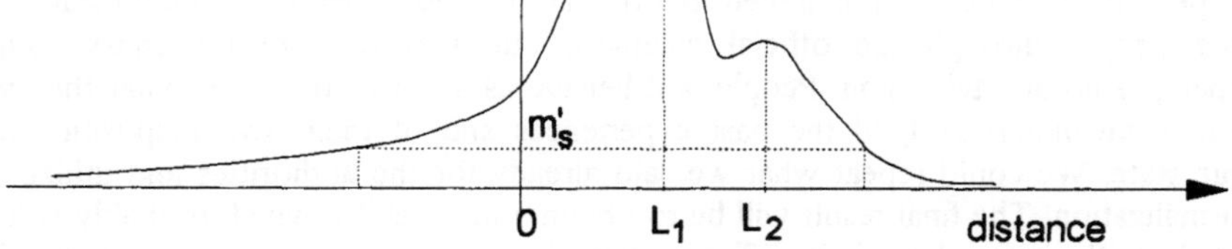

Fig. 5

In fig. 5 the maxima are maximised because to the r_i have been attributed the (arbitrary) values deduced from fig. 4 (they can be > 1). There are no difficulties, of course, to normalise also the values of the r_i, imposing the condition: $0 < r_i < 1$. In such a case the maxima will be less emphasized.

In the case of S_3 too, we can establish a threshold value m_s', below which the points are not elements of S_3. So doing, it is clear that now S_2 and S_3 are no longer coincident, but only partially overlapping. S_2 and S_3, on the basis of the distributions of figures 3 and 5, are schematically represented in fig. 6, arbitrarily looked down on a plane.

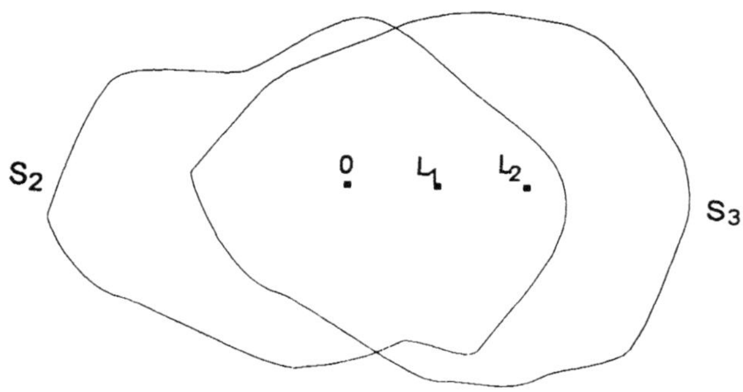

Fig. 6

In the best case, then, the preventive necessary measures will be assumed only with respect to the area corresponding to the intersection of the two sets. Also in this area the intensity of the assumed measures will not be that implicitly suggested by the predictions of the scientists. In other words, for a given point x measures of security will be taken if, and only if $X \in (S_2 \wedge S_3)$. We can only hope that the intersection of S_2 and S_3 is always very large, although the presented considerations cannot guarantee for that.

As we have said and indicated in fig. 1, there is a last step to establish what will really interest people. The decision assumed by the authorities must be communicated to interested people through the official channels, but also through the mass media: newspapers, radio and television. People will behave as a consequence of what they will believe to have understood, of the past experiences and of their own disposition and economic state. We could repeat what we said already for the authorities and utilize the same formalization. The final result will be to obtain a new set S_4, which probably will be different from S_2, more than S_3 is different from S_2. We could obtain a schematization similar to that of fig. 6, but with a smaller area of overlapping. Additionally, this situation will be made worse by two facts:

a) the direct channel presumably acting between authorities and scientists (they can explain and clarify their own opinions), does not exist, or exists only partially, in the case of public opinion;
b) on the effective behaviour of people strongly play a role the past experiences.

The last consideration puts in evidence a further problem, acting also in the case of the authorities behaviour. A system like that described in fig. 1 is characterized by a strong memory of the past, differing, in this respect, from every good statistical system. A not verified prediction will carry to underestimate a new prediction, while a verified prediction will carry to overestimate the new one. There is the possibility that people do not make what it has to be done or that dramatize situations of moderated risk.

To conclude: in this field, scientists have great responsibility as it concerns the society in which they live, but they risk to be unheard or misunderstood. Erroneous decision could be assumed in these situations. In these cases it becomes important a problem of communication. Scientists, especially in places at risk, will have to stimulate a generalized information. But this is a long and difficult work. The scientists have also to find out a way to translate their difficult scientific language in a simpler one. But the scientists have also the difficult problem to be seriously listened by authorities and public opinion. Is it a problem of "dramatization"? Of choosing the appropriate tactics or strategies? Or of what else? I cannot discuss this argument, also because it asks for competences I do not have and which could be found in the specialists of communication theory, in the sociologists, in the specialists of language, may be in the specialists of advertising or in the artists. But in the last cases what are the conditions to avoid that communication becomes a source of further distortion?

REFERENCE

Mandani, E.M., Gaines, B.R. (eds.), (1981) "Fuzzy Reasoning and its Applications". Academic press, London.

UNCERTAINTIES IN NATURAL HAZARDS PREDICTION AND ITS EFFECT ON USER COMMUNITIES PERCEPTION: SOVIET UNION CASE STUDY

BORIS N. PORFIRIEV[1]
Institute for Systems Analysis
Russian Academy of Science
9 Prosp 60 Let Octiabria 117 312
Moscow
Russia

INTRODUCTION

There is a current opinion that modern civilization has reduced the risk of natural hazards. A thorough analysis of empirical data gathered over the years, however, argues that our society continues to be highly vulnerable to natural disasters that are more destructive then ever. My estimate shows that average losses inflicted by these disasters in the former Soviet Union have now exceeded two percent of GDP per year. Therefore, economic and intellectual input into forecasting should be increased. More accurate and sophisticated forecasting would reduce the number of victims and their losses from natural disasters.

Factors of uncertainty in natural hazards forecasting

Accurately forecasting a disasters' place, time and destructive force is difficult. Complications are primarily predefined by complexity of the possible disaster that includes a variety of changing factors. The situation is further aggravated by the additional anthropogenic factors that add to and reinforce the traditional and purely natural causes of disasters. The latter dramatically increases the volume of input data necessary for prediction. In general, these complications reflect the limits of modern knowledge and technicnology, though the progress in instrumental prediction in leading western countries is impressive. The situation in the Soviet Union is quite different and will be discussed later.

Uncertainty in prediction of the natural hazards main parameters is inherent in forecasts. The latter plays a vital role in providing timely protection for lives, health and properties of the communities located in hazard prone areas. An ironic paradox is that the highest

[1] Institute for Systems Studies, USSR Academy of Sciences

J. Nemec et al. (eds.), Prediction and Perception of Natural Hazards, 49–53.

uncertainty and, consequently, the least accuracy are intrinsic to this type of prediction. The result is serious problems for decision makers using the results and inhomogeneity in natural hazard perception by user communities.

Types and modes of user communities perception

Practical experience and theoretical research permit the conditionall delineation of two main types of perception and reaction induced by uncertainty in natural hazard prediction. These are the radical (extreme) and the rational (moderate). The RADICAL type embraces two perception modes, one may be characterized as indifferent, passive and even fatalistic perception. This mode, like the radical type in general, is more closely linked to the uncertainty of natural hazards prediction than is the rational type. Applying too indifferent or fatalistic a perception mode, lay people view this as ignorance, charlatanism and wild guesses. The forecasters and their assessments are taken as irrelevant and nontrustworthy information.

Such a perception means passiveness in terms of reaction, including unwillingness of community members to insure their lives and property, and the inclination of local authorities to waste time on preparing plans to defend communities jeopardized by natural disasters. In the USSR laws, legal enforcement and administrative control in this specific sphere are either lacking or are poor. This is the reason these plans are often a formality or do not exist at all.

Obviously, the explanation of such a perception mode and manner of action goes far beyond prediction's stochastic character. No less important are the other reasons, particularly socio-psychological factors. Russian society for decades has lived with the expectation of disaster. Therefore, when a natural hazard or another emergency prediction occurs, user communities may perceive it as nothing more than additional evidence or confirmation of their own grim moods and misgivings. There is a constant expectation that the situation will become worse, no matter what the specific situation. This environment is saturated with social apathy and the typical Russian hope of the off chance; with the deformed social relations and economic structures that encourage the birth and preservation of vicious labor morals; with the priority of current and individualistic interests over perspective ones and the interests of social groups and society as a whole. As a logical result, prediction warnings are neglected.

The consequences may be vividly illustrated by the case of the Spitak quake in 1988. The Armenian Republic is a seismic prone region where as early as 1985 specialists detected a forecast anomaly. This became the site of a later tragedy. In spite of this knowledge, when erecting schools, houses, etc., construction workers in Leninakan and other cities used inferior materials and ignored established procedures. In doing so they prepared the graves for themselves, their children and other community members. Ironically, those who failed in their responsibility to warn the community of the impending earthquake (specialists from the Leninakan Seismological Institute) perished in the disaster.

Discussing this tragedy at UNESCO seminar, "Spitak-88" held in 1989 in Erevan, some international experts considered such an indifference to hazard as inexplicable. I believe that the above reasoning concerning the specifics of social environment of the future calamity explains, at least partially, its deep roots and appropriateness.

There is another perception mode by the user communities inside the radical type that may be described as over-active or even apocalyptic. The logic, reasoning and behavior in this situation differ dramatically from the indifferent or fatal perception mode. The people whose typical perception and reaction are apocalyptic and, in case of emergency, panic, consider forecast uncertainty as under-assessment or real risk. They often point to the grim examples of Spitak or Ashkhabad in 1948, where the possibility of a great quake was ignored by the specialists. In the framework of the over-reactive perception mode, concrete responses vary greatly. At one end of the range is a trust in the prediction's ability to foresee the disaster event, although that view is not accurate enough to define the parameters. At the other end is a complete disbelief in the ability of modern science to predict the consequences. From the later response comes a reliance on fortune-tellers, prophetesses, etc., for information perceived to be more accurate. The over-reactive perception is aggravated by poor and often unverified media communications, encouraging the community to over-react. Recently in some of the soviet cities media issued predictions made by prophetesses fixing the exact place and time of the future quake. Despite the professional's opinion publicly disproving these forecasts, thousands of people left their workplaces and homes. Naturally, the disaster did not take place and communities suffered substantial economic losses and nervousness. Nevertheless, some inhabitants continue to consider themselves lucky to escape death and thus justify their over-vigilance.

To my mind, the over-reactive mode effect can also be traced in the manner of explaining causes of potential or actual natural disasters that are perceived by certain social groups or people as a result of a plot. In December, 1988, even the most intelligent circles of the Armenian society supported the idea of a preplanned explosion aimed at distracting communities' attention from the Kharabakh conflict and designed to provoke an earthquake in Spitak region.

Unlike the radical or extreme perception, the RATIONAL or MODERATE type is inclined to actively defend against predicted natural hazard. The degree of uncertainty of the forecast is only one factor influencing preparation and training. By itself, however, the rational mode is incapable of changing the behavior mode. This type of perception relies upon trustworthy risk communication, existence of initiative, wise persons serving as a nuclei or growth center for emerging volunteer's groups and getting active support from local authorities. In this case, uncertainty in natural hazard prediction no longer continues to be an inhibitor for human emotions. Rather, it performs signal functions for accelerating active response measures to meet possible disaster. It also plays a testing role in checking the preparation and mobilization potential of a community with negative experiences in previous calamities. The latter often are perceived as images of punishment for neglecting nature's powerful forces and as a lesson for the future. Simultaneously, the predicted disaster takes the image of God's trial that should be overcome with dignity. Examples of this mode of perception are well known everywhere, including the Soviet Union.

At the same time, the rational perception mode does not necessarily lead to adequate reaction and action modes in case of an actual natural disaster. The trigger depends on various conditions including socio-economic, political, cultural, historical, etc., factors. I will limit further discussion to brief comments on the first of these factors.

Deficiencies in financial and material resources and poor technical equipment are typical of the current and transitional state of Soviet society. Improvements in these areas are needed to protect hazard prone areas. These deficits reduce the positive effects of the rational perception and reaction mode of the communities.

The case of Kamchatka Peninsula in the Soviet Far East reveals that about 60 percent of the houses do not correspond to the construction code prescriptions on seismity. These factors mean that local communities lack time and resources for the necessary repair and reinforcement to prepare for the anticipated major earthquake. Appeals to Moscow resulted only in vast supplies of "crows." This can hardly be taken as an effective mitigating instrument. There are communities in this country of volcanoes that are well familiar with quakes. They usually perceive uncertainty in prediction as quite natural. They evacuated to the mainland several times last year after experiencing the grim forecasts and feeling defenseless against terrible danger. The only support they counted and may rely upon can be provided by special rescue commands of the Ministry of Interior's Fire Division and by volunteer groups from mainland. Realization of this fact burdens the local communities' morale.

Conclusion

The situation in the Soviet Union briefly discussed in this paper should and can be changed only in the context of radical reforms in politics, economy, social and cultural areas started by perestroika. The morale of lay persons and whole communities, including those in hazard prone areas, primarily depends upon and are closely linked with social stability and economic progress. This link is in terms of both ethnic conflict prevention and adequate food and consumer goods supplies that are perceived by soviet citizens as more urgent and important than hazards induced by natural forces even in the mentioned areas. The success of perestroika would include creation of a favorable political environment and socio-economic prerequisites for realizing rapid and effective actions to defend communities living in vulnerable natural milieu. It would also mean improvement in morale through sound risk communication and training, thus making the perception of all uncertainties and dangers connected with calamities more rational and reasonable and the reaction more active and effective.

REFERENCES

Journal Articles

Arabov, V., Kazaryan, A., Porfiriev, B., Prigozine, A. and Kharash, A. (1989) "Disaster: Lessons and Hopes." Znanie-Syla, 6, 7 (in Russian).

Drabek, Thomas E. (1983) "Alternative Patterns of Decision-Making in Emergent Disaster Response Networks." International Journal of Mass Emergencies and Disasters, 1, 2 August, 277-305.

Dynes, Russell R. and Quarantelli, E. L. (1976) "Community Conflict: Its Absence and Its Presence in Natural Disasters." Mass Emergencies, 1:139-152.

Mileti, Dennis S. (1983) "Societal Comparisons of Organizational Response to Earthquake Predictions: Japan vs. United States." International Journal of Mass Emergencies and Disasters, 1, 3: 399-414.

Nigg, Joanne M. (1982) "Communication Under Conditions of Uncertainty: Understanding Earthquake Forecasting." Journal of Communication, 32:27-36.

Sobolev, G. A. (1989) "Problems of Quakes Prediction." Priroda, 12:47-54, (in Russian).

Turner, Ralph H. (1983) "Waiting for Disaster: Changing Reactions to Earthquake Forecasts in Southern California." International Journal of Mass Emergencies and Disasters, 1, 2: 307-334.

Book references

Quarantelli, E. L. (1978) Editor. Disasters: Theory and Research, Beverly Hills: Sage Publications.

Preliminary papers

Pelanda, C. (1981) "Disaster and Socioeconomic Vulnerability." Preliminary Paper 68, Columbus, OH: Disaster Research Center, The Ohio State University.

SESSION 2

UNCERTAINTIES IN THE DEVELOPMENT OF PREDICTIONS OF LARGE SCALE ATMOSPHERIC PHENOMENA: DROUGHTS, TORNADOES AND HURRICANES

UNCERTAINTIES IN THE DEVELOPMENT OF PREDICTIONS OF LARGE SCALE ATMOSPHERIC PHENOMENA, DROUGHTS, TORNADOES AND HURRICANES

J.L. RASMUSSEN
World Meteorological Organization
Case Postale 2300
CH1211 Geneva 2
Switzerland

INTRODUCTION

This overview paper addresses the issue of the uncertainties we face in the predictions - for purposes of natural disaster reduction - of atmospheric phenomena ranging from large scale extra-tropical storms and droughts to tropical cyclones (hurricanes and typhoons) and tornadoes. I will address each phenomena summarizing the prediction capabilities today and, where possible, give some idea of the outlook for the future.

The ability to predict atmospheric phenomena depends upon the following five factors:

- Physical understanding of the phenomena, the model
- Observations (data base) available for the analysis
- Sophistication of the analysis and initialization scheme for utilizing the observations
- Horizontal and vertical resolution of the model
- The model's ability to describe or replicate the physics of the atmosphere including subgrid scale processes.

Of course, the five factors must be considered differently for each of the widely different scales (both space and time) characteristic of each of the phenomena addressed in this paper. The intent in the following sections of this short paper is to provide just an overview sketch of the issue of uncertainties in order to provide a background for discussion.

Large Scale Weather Prediction

The major progress made in meteorology over the past several decades is in the development of global numerical models and their application to large scale weather prediction. For example, during the 1980's at the World Meteorological Center (WMC),

J. Nemec et al. (eds.), Prediction and Perception of Natural Hazards, 57–63.

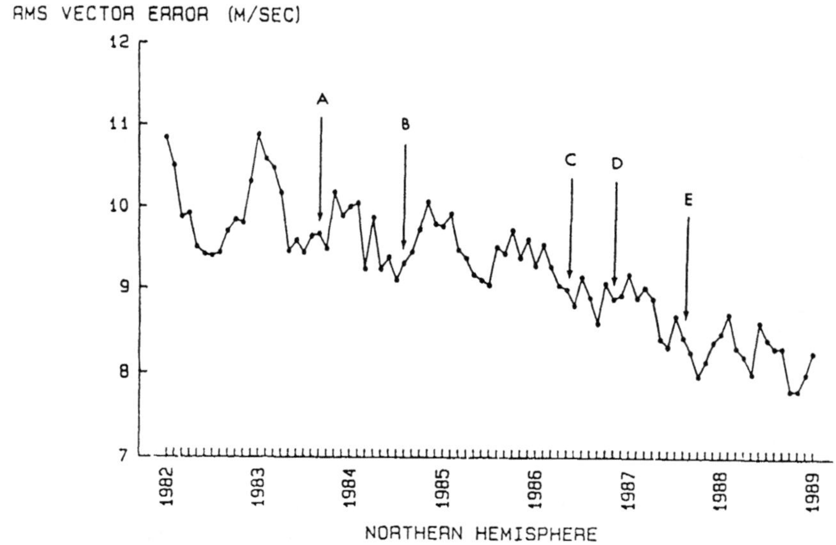

Fig. 1: 250mb RMS vector error for NMC operational 24-hour forecasts (January 1982 through January 1989)

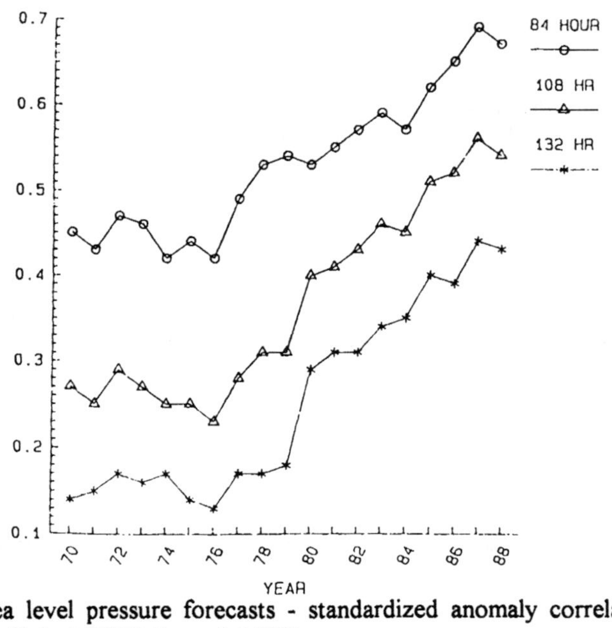

Fig. 2: Sea level pressure forecasts - standardized anomaly correlation coefficient (1970 through 1988)

Washington, D.C. the root-mean-square-error for twenty-four hour forecast of winds at 250 hPa (approximately jet-stream level) has decreased from 10 m/s to 8 m/s. Considering the estimated error in the analysis is approximately 5 m/s, this change is indeed dramatic. This statistic is illustrated in Figure 1 (Bonner, 1989) . The arrows labelled A through E indicate dates when major changes in the made or computer facilities were incorporated. The two most recent improvements were the introduction of 18 layer full physics Model (D) and improvement in horizontal resolution to the equivalent of 160 km (E).

The current level of accuracy of WMC Washington predictions is illustrated in Figure 2. This shows the record of skill of the 84, 108 and 132 hour numeriCal sea level pressure forecasts. The general trend in skill since 1976 (when a seven layer primitive equation model was operationally implemented) has shown a systematic improvement with time as the model and the computing facilities have been systematically improved. The 132 hour forecasts are now roughly as accurate as the 84 hour forecast was ten years ago.

Experts from the major modelling centres now feel that to continue to make major advances in prediction depends on improved data coverage. This is especially true for the data void areas of the globe which include the oceans, especially the southern hemisphere oceans, the African and South American continents and the polar regions.

Most severe extratropical cyclones are simulated in the numerical forecasts now. Explosive development of fast moving storms over the ocean still are occasionally a surprise but given adequate observations even this problem can be solved.

Droughts

Droughts arise from a complex interaction of many factors with land-use and water management activities, for example, contributing to the precipitation deficit and evaporation excess leading to the drought condition. Numerical model. predictions of the onset. of drought are not yet possible. Indeed the ability to predict weather patterns even in a large space scale for beyond 10 days is not considered possible. Once a drought is established the numerical models can be useful for predictions of the continuance or break in the drought ever the next 5 - 10 days. There is empirical evidence of a relationship between the ocean-atmosphere system and the distribution of temperature and precipitation, the so called tele-connections. One example is the relationship of the El Nino-Southern Oscillation phenomena in the Pacific Ocean and the subsequent precipitation distribution over widely varying regions of the globe as illustrated in Figure 3 (Ropelewski and Halpert, 1989). Major research projects are underway to understand the phenomenon which will hopefully lead the way to a predictive capability. There presently is little skill in the 30 to l00 day lead time prediction schemes that have been tested.

Tropical Cyclones

With the advent of geostationary meteorological satellites in the mid-l970's, the capability

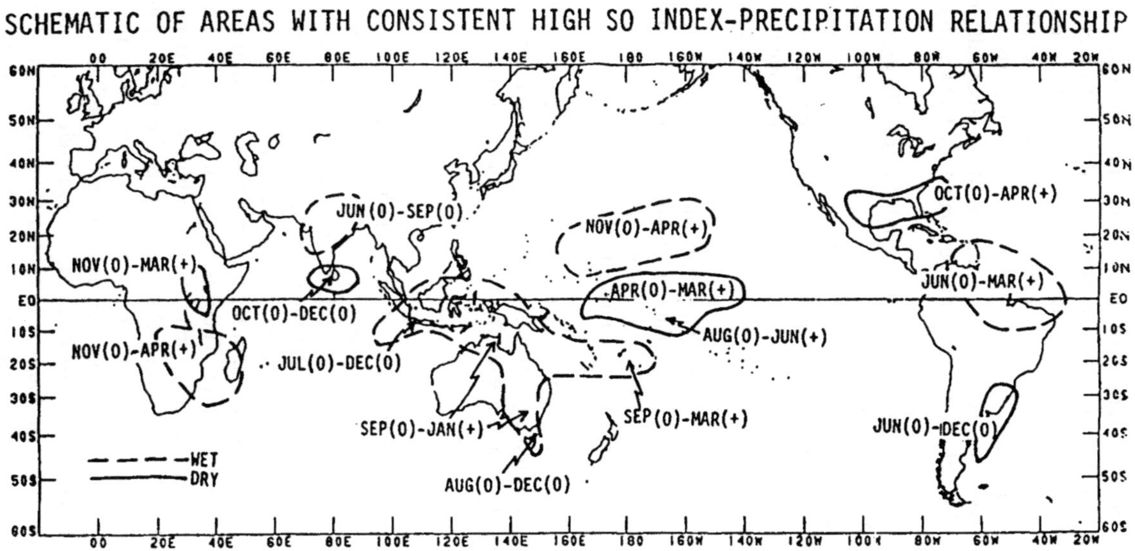

Fig. 3: Schematic representation of the principal areas of high-SO index related precipitation based on the analysis of precipitation composites and time series

to monitor and track tropical cyclones has made a major difference in the predictions of these systems. The numerical models described above for large scale weather prediction now have spatial resolution that accommodates the tropical cyclone scale and these have proved useful in the tropical cyclone warning programmes The requirements for tropical cyclone predictions are forecast of the track, time of landfall and the intensity of the storm. The major disaster with the tropical cyclone is usually caused by the storm surge driven by the winds, the flooding that often is triggered as the storm moves inland and, of course, damage caused by the wind. The predictions required therefore consist of a 'package" of forecasts that include storm surge and flood forecasts, as well as intensity and track of the storm.

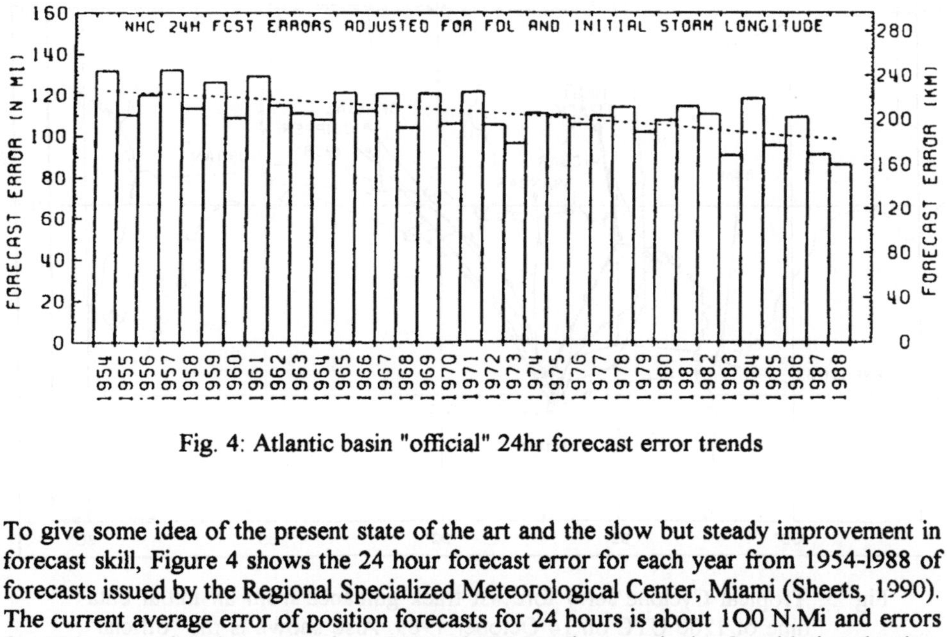

Fig. 4: Atlantic basin "official" 24hr forecast error trends

To give some idea of the present state of the art and the slow but steady improvement in forecast skill, Figure 4 shows the 24 hour forecast error for each year from 1954-1988 of forecasts issued by the Regional Specialized Meteorological Center, Miami (Sheets, 1990). The current average error of position forecasts for 24 hours is about 1O0 N.Mi and errors for 48 hours and 72 hours are about 220 and 370 N.Mi respectively. Considering the time needed to evacuate some coastal communities this error means that a larger section of coast must be warned than will ultimately experience the storm with the subsequent "false alarm" problem for most of the coast warned.

Figure 5 illustrates the difficulty in prediction of the track of a weak tropical cyclone imbedded in a weak steering current when the various forecast techniques provide a wide variety of guidance. Most forecast cases are more clear-cut than this one, but I use this illustration to demonstrate the difficulty of the problem.

Tropical cyclone experts expect a continual gradual improvement in the quality of predictions. The use of specialized 'imbedded' mesoscale numerical models that provide a much more detailed computation grid for areas within the global model grid (e.g. for a tropical cyclone and surroundings) will provide the next step. Better data coverage over the tropical oceans is also expected to pay rich dividends in cyclone prediction.

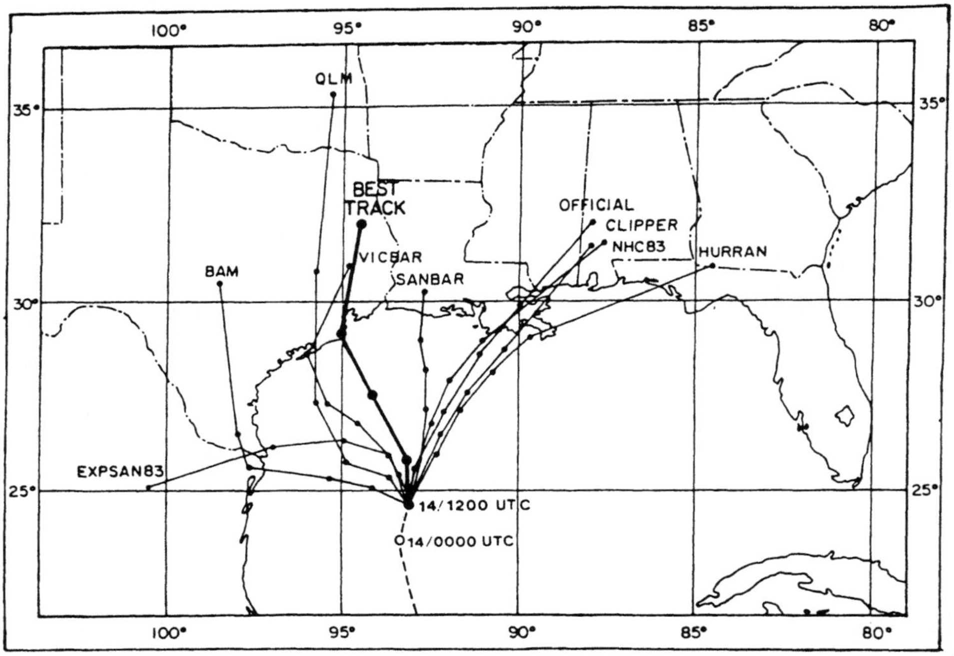

Fig. 5: Tropical Cyclone Jerry forecast track guidance from an initial data time of 1200 UTC on 14 October 1989. Also shown is the "official" forecat track issued at the same time and the actual track of the tropical cyclone (Sheets, 1990)

Tornadoes

A prediction capability for the small scale violent phenomena of tornadoes does not exist. Based on large-scale numerical weather prediction analysis and forecasts information providing an outlook for severe thunderstorms and possible tornadoes with a six to twelve hour lead time is possible. However, the area covered by these outlooks mast be very large in comparison to the actual weather system that evolves and, therefore, are not effective warnings.

Current tornado warning schemes are based upon the detection of the phenomena as early as possible, observing the rotation and other cloud characteristics prior to funnel cloud formation, and the immediate issuance of a warming based on the hard evidence. New technologies, including doppler radar systems and improved satellite observations are contributing to this detection capability. The objective of the warning schemes is to provide warnings with about 20 minutes lead time and to as small a risk area as possible.

The development of operational numerical models directly applicable to the tornado prediction problem are only in the development stage at this time. The doppler radar networks being implemented will contribute data bases to be used in this research effort.

Concluding comment

The purpose of this paper has been to summarize a few salient points which can be used as a basis for discussion- The paper does not touch on a critical link in the chain of activities leading to natural disaster reduction - that link is the delivery of the predictive information to the community in an effective way which results in the appropriate response or reaction on the part of the recipient. The other papers presented at this workshop will contribute to building that link.

References

Bonner, W.D. 1989: NMC Overview: Recent Progress and Future Plans, Weather Forecasting 4: 275-285.

Miyakoda, K., G. Hembree, R. Strickler and I. Shulman, 1972; Cumulative Results of Extended Forecast Experiments. 1: Model performance for Winter Cases,.Mon.Wea.Rev., 100: 836-855.

Ropelewski, C.F., M.S. Halpert, 1989: Precipitation Patterns Associated with the High Index Phase of the Southern Oscillation, Journal of Climate, 2:, 268-264.

Sheets, R.C,, 1990: The National Hurricane center - Past, Present and Future, Weather Forecasting 5: 185-232.

EMPIRICAL STUDIES OF PUBLIC RESPONSE TO TORNADO AND HURRICANE WARNINGS IN THE UNITED STATES

EARL J. BAKER
Department of Geography
Florida State University
Tallahassee, FL 32306-2050
U.S.A.

TORNADOES

Response levels

Three studies (Hanson, Vitek, and Hanson, 1979; Glass et al., 1980; Hodler, 1982), and to a lesser degree a fourth (Taylor, Zurcher, and Key, 1970), provide most of the survey data documenting public response to tornado warnings.

Hanson, Vitek, and Hanson conducted their survey in Flint, Michigan site of a very strong tornado more than twenty years earlier. There was no specific tornado warning prior to the event, but respondents were asked in general what they did when they heard a tornado warning. Twenty-seven percent said they took shelter immediately, 48 percent said they checked the weather to determine if they were at risk, and 25 percent said they continued doing whatever they had been doing.

Glass et al. conducted their survey following a strong tornado in Wichita Falls, Texas, which killed 47 people. Seventy seven percent said they took shelter before the tornado struck. Fifty three percent of the respondents living in single-family homes went to interior closets, bathrooms, and hallways, and 24 percent went to basements or storm cellars. Sirens were sounded at least twice in the area as much as an hour before the tornado arrived. One of the most widely cited findings of the study was that a significant number of residents left their single-family homes and tried to evade the tornado in their cars, resulting in a higher than normal incidence of deaths in autos. Officials now explicitly advise against that response.

Hodler interviewed residents of Kalamazoo, Michigan and found that 48 percent said they went to shelters immediately upon hearing sirens, 22 percent attempted to verify the threat, and 18 percent ignored the warnings. Seventeen percent said they had no idea what the sirens signified.

J. Nemec et al. (eds.), Prediction and Perception of Natural Hazards, 65–73.

With a less random set of interviews, Taylor, Zurcher, and Key reported less quantitatively precise findings after a tornado in Topeka, Kansas in 1966. Most people took shelter upon hearing sirens, even though they doubted that it was necessary. Many residents also indicated they normally did not take shelter so readily, but "somehow that day was different."

Predictors

Experience is not a good predictor of response. The main point of Hanson, Vitek, and Hanson, was that awareness of the 1953 devastating tornado was a better predictor of response than was personal experience in tornadoes. Response did not vary at all with previous tornado experience. Hodler did not report response by experience or awareness level but did measure a number of awareness indicators and found awareness to be generally poor. Eighty-one percent of the respondents said they had a plan before the warning to go to a basement or other shelter; and 93 percent did so. Those who did not know the meaning of the sirens were less likely to take shelter immediately.

Taylor, Zurcher, and Key suggested that a number of factors might have caused people in Topeka to take the warnings more seriously than was typical. The sky was an unusual color, many had heard that another tornado struck a Kansas town earlier in the day, radio announcers reading alert scripts sounded urgent and concerned, and there was electrical static over the radio.

One study by Sims and Baumann, (Sims and Baumann, 1972) that did not employ post-warning survey data has been one of the most frequently cited. This is partly because it appeared in *Science* and partly because of its controversial conclusions. The authors noted that the number of deaths per tornado occurrence is greater in southern states of the United States than in northern states. They were unable to account for the difference with considerations such as differences in tornado severity, quality of building construction, and type of warning system. Upon administering a psychological instrument to measure locus of control in a southern state (Alabama) and in a northern state (Illinois), the authors found that Southerners were more fatalistic and less likely to believe they could control their destinies than did Northerners. Sims and Baumann suggested that Southerners were, therefore, less likely to take less adaptive action in response to tornado warnings. This could account for their higher fatality rate. A flood of letters to the editor offered an alternative explanation. The explaination was based on the greater number of homes in the North that have basements (the best protection in a tornado) than in the South.

HURRICANES - Evacuation rates

Evacuation rates vary from place to place in the same hurricane and from storm to storm in the same place. It is clear that in some locations almost everyone leaves while in other locations within the warning area half or less of the population at risk evacuates. Studies reporting on evacuation behavior in a single, small geographical area or that report one figure for a large area can be misleading. The concern that "people won't evacuate.", is erroneous if it means that most people won't leave most of the time they should.

Less obvious might be the variation in response which can occur from threat to threat in the same location. In Galveston 68 percent left during Carla in 1961, 63 percent during Allen in 1980, and 47 percent during Alicia in 1983. In Panama City Beach, 87 percent evacuated during Eloise in 1975 but only 54 percent druing Frederic in 1979. However, 80 percent evacuated in Pass Christian, Mississippi during Camille in 1969 and Frederic, ten years later. These figures too, caution against generalizing from a single evacuation. The threat posed by all storms and people's information about them are not the same; so neither are public responses.

Predictors (and nonpredictors) of response

Risk area: Low-lying sites exposed to open water are more dangerous than sites at higher elevations even a few city blocks inland and protected by offshore barriers. Typically high-risk sites are barrier islands or open coastlines with elevations less than 10 feet above mean sea level. Low-risk sites are usually on the relatively protected mainland, over 15 or 20 feet in elevation. In between are moderate-risk sites, on the protected mainland but no more than 10 to 15 feet high and often flood-prone in strong hurricanes. Fortunately, evacuation from high-risk areas is usually very good. When officials are successful in notifying residents of the areas that they need to leave, more than 90 percent usually comply.

Emergency preparedness planners have traditionally been understandably preoccupied with maximizing evacuation rates from high-risk areas. Moderate-risk areas, while certainly safer than high-risk locations, are also subject to inundation by storm surge, and evacuation rates in these areas is usually between 55 and 65 percent. It appears that moderate-risk areas deserve more attention from officials than they have previously received.

Evacuation rates in low-risk areas pose a different sort of dilemma. The great majority of residents in these locations would be safe in their homes in all but the most severe hurricanes (the most notable exceptions being mobile-home dwellers). Roughly a third, however, evacuate, contributing to traffic congestion. They increase the time necessary to clear evacuees from high- and moderate-risk areas, time which could well be insufficient. Advising low-risk area residents to stay put throughout a threat, however, is a legally risky and ethically uncertain proposition.

Evacuation notices: Advice or orders from public officials affect evacuation rates more than any other factor, with the possible exception of risk area. The two variables, though, are confounded. Two illustrations are compelling. In Mobile during Frederic, the evacuation rate was only 34 percent in an area closely corresponding to the 100-year floodprone area (a moderate-risk location according to the trichotomy set forth above). In the areas where evacuation was advised, however, 63 percent evacuated. Of the Mobile residents who said they heard from public officials that they should evacuate, 84 percent left; of those who didn't hear, the evacuation rate was only 20 percent. In Miami and Miami Beach during David, the difference was even more pronounced. Eighty-eight percent of those who heard they should leave evacuated, while only eight percent of those who didn't hear did so. Virtually all of the sample was in high and moderate-risk areas where evacuation was ordered by state and local officials.

The distinction between evacuation recommendations and evacuation orders is not as important as many public officials and preparedness professionals believe. Orders are by far the less common of the two. Although most residents who hear recommendations understand that they are advisory, 15 to 25 percent often interpret them as orders. More important than the legal difference between an order and a recommendation is the manner in which the notice is disseminated. Residents need to understand without question that an evacuation notice applies to them, and more personalized modes of delivering the message result in higher responses. In the Mobile and Miami examples referenced above, officials issued evacuation advisories and orders, but many residents either didn't hear the notices or didn't understand that the message applied to them.

Threat factors: Expectation of damage is a better predictor of evacuation than belief the storm will hit one's location. Residents might be convinced that a storm will strike their area, but believe it is too weak to pose a threat to them. People who believe that winds will be strong enough to overturn autos or that their homes will suffer water damage are more likely to evacuate than people who don't.

The National Hurricane Center issues a hurricane *watch* when a storm is expected to make landfall within the next 36 hours. It issues a *warning* when the storm is believed to be within 24 hours of land. Watches and warnings alone do not prompt significant evacuation. Evidence will be presented below when evacuation timing is discussed.

Hurricane experience: Hurricane experience is a difficult concept to define and measure, although some cases are clearer than others. The conventional wisdom is that people in communities that have experienced major hurricanes "recently" will evacuate in greater numbers than people in communities that have not. People in communities without recent major hurricane experience, but have been on the fringes of a bad storm or experienced a lesser hurricane are perceived as more reluctant to evacuate. Those people are believed to have "false experience." Windham et al. (1977) coined the term "experience-adjustment paradox" in reference to the belief that newcomers to coastal areas would actually be more likely to evacuate than oldtimers because the newer residents had no storm experience, true or false.

There is no consistent relationship between experience and evacuation. Part of the reason might stem from the difficulty in defining and then measuring experience. Early empirical support for the experience-response connection came from Hurricane Carla. Cameron Parish, Louisiana had the highest evacuation rate (96 percent) of any location surveyed after the storm, although Cameron was on the edge of the warning area. Hurricane Audrey, which killed 400 Cameron Parish residents in 1957, was credited with conditioning the residents to evacuate. It is also true, however, that Cameron Parish was the only survey area during Carla where local officials ordered residents to evacuate. The Audrey effect might not be as obvious as many thought (except perhaps indirectly, by inducing local officials to issue the evacuation order).

Ninety-seven percent of Pensacola and Pensacola Beach evacuated before hurricane Frederic in 1979, although Pensacola had no recent major hurricane history. There were ample opportunities for "false experience" in near-misses from Camille in 1969 and Eloise in 1975. Evacuation in Frederic was higher in Pensacola than in Panama City Beach which was also in the warning area and suffered major damage from Eloise. Pensacola's evacuation was also greater than Pass Christian, Mississippi's during Frederic, although Pass Christian was been ravaged by Camille 10 years earlier. Grand Isle had an evacuation rate comparable to Pensacola's during Frederic, despite Grand Isle's location on the edge of the warning area. Also, Hurricane Bob, a very weak storm, struck the island earlier in the summer, providing a golden opportunity for residents to gain "false experience."

This is not to suggest that people don't have false experience. Many certainly do. That is, many people incorrectly think they have been through major hurricane conditions. (Leik, Carter, and Clark, 1981) The issue, though, is whether those or any other hurricane experiences have a predictable effect upon subsequent evacuation behavior. People have been asked if they had any kind of previous disaster experience, any kind of previous hurricane experience, how many hurricanes they had experienced, how recently they had experienced a hurricane, how severe was the hurricane they experienced, how much property damage they had experienced in hurricanes, and whether any family member had ever been injured in a hurricane. Most research has relied upon self-reports of experience. Some has gone farther and recorded where people said they were living in different years and then assigned experience levels based upon meteorological records. None of the measures are generally predictive of what residents did in subsequent hurricanes.

In some evacuations, where data was collected only in high-risk areas, the overall evacuation rate was so high that there was too little variance in the response for any meaningful test of statistical significance. These cases too could be viewed as refuting the experience-response hypothesis or any other predictor hypothesis. The variance was so small because almost everyone in the sample responded the same way, regardless of differences in experience or whatever else. Even if larger samples had produced statistically significant differences in response, the magnitudes would almost certainly have been trivial.

The main reason to expect that experience might affect evacuation behavior is that experience contributes to awareness of the hazard, usually meaning an appreciation of the danger posed by it. There are many ways to achieve that kind of awareness, some of which might lead to greater concern than experience.

Length of residence: Closely tied to the experience-response hypothesis is the notion that people's evacuation behavior depends upon how long they have lived in the coastal area. There are two versions of the idea. The first is that newcomers don't appreciate the destructive potential of hurricanes or know what to do. Consequently, they are less likely to evacuate than people who have lived there longer and learned to respect the hazard. The second is that very few coastal areas have experienced direct hits by major hurricanes in the past five or ten years. Therefore, people living in coastal areas at least that long often believe they have experienced such storms and found them not so dangerous (i.e., "false experience"). Therefore, newer residents are actually the more likely group to leave.

There is no consistent evidence for either, although length of residence has been measured and tested against evacuation in several hurricane studies. Even Windham et al.'s claim of support during Eloise for the "experience-adjustment paradox" isn't upheld by a reanalysis of their own data (Baker, 1979).

Hurricane awareness: It might seem that people who know more about hurricanes are more likely to evacuate than others, but they don't. Studies have measured the number of National Weather Service hurricane safety rules people can name, number of hurricane terms known, whether people have ever seen a copy of the rules, whether they possessed a copy of the rules, whether they have ever even heard of the rules, whether they know the difference between a watch and warning, and so forth. People who know more are no more or less likely to evacuate than people who know less. It appears that there are people who know a great deal about hurricanes but are unlikely to evacuate, while there are other people who know hardly anything about hurricanes but might be terrified by them.

There is, however, evidence that if people realize that their homesite is hazardous and subject to inundation by storm surge, they are much more likely to evacuate. Leik, Carter, and Clark suggest that warnings and even generally worded evacuation notices only confirm residents' perceptions of their own vulnerability (1981). Thus, people who believe their homesites are safe won't interpret the warning or evacuation notice as applicable to themselves. Printing hurricane safety rules and tracking charts on grocery bags can do no harm, but if public awareness programs are to significantly enhance evacuation they must concentrate on communicating to coastal residents their personal risk.

Crying wolf: Preparedness officials frequently express concern that if people evacuate unnecessarily during a hurricane threat, they won't be willing to leave during the next threat. Although this is a difficult hypothesis to test, there is no evidence to suggest that it creates a major problem. As early as 1954 evacuees who left Panama City in a "false alarm" were asked whether they would be more reluctant to leave in the future, and few said they would. However, many stayers said they would leave in the future (Killian, 1954).

In 1985, residents of Panama City, Florida were told on three separate occasions to evacuate from threatening hurricanes. The storms missed all three times. There was no discernible reduction in response across the three evacuations.

Housing: Mobile home residents of coastal areas are usually advised to evacuate even if they live outside the areas expected to be inundated by storm surge. They appear to be about twice as likely to evacuate as other residents of comparable risk areas, but sample sizes in most evacuations have been too small to draw consistent conclusions. Sample size has also been a problem in studying responses of residents of multi-story buildings, but existing evidence suggests that residents of such structures in high-risk appear to evacuate in roughly the same proportions as other residents when told to do so by officials.

Miscellaneous variables: One of the better things to know in predicting whether someone evacuated is the extent to which the *neighborhood* evacuated. If most of the neighborhood evacuates, a resident is more likely to leave than someone in a neighborhood where most people stayed. The traditional interpretation of this finding is that there is a conformity effect. That is, one resident leaves (or doesn't) because the neighbors are leaving (or aren't). This no doubt affects some people, but neighborhood response is badly confounded with risk area and often with actions by public officials. A resident doesn't necessarily leave or stay because neighbors engaged in the behavior. Often individuals and neighbors are leaving or staying for the same reasons (the neighborhood is subject to flooding, police have gone door-to-door advising evacuation, and so forth). Although the conformity effect probably exists, it is unlikely to be as strong as often claimed.

Demographic variables such as age, income, and race have not generally been found to predict evacuation. Boat ownership, church attendance, pet ownership, operating home weather instruments, and total number of emergency preparations taken are some of the other variables tested and not found to be consistent or strong predictors of response.

References

Journal Articles

Baker, Earl J. (1979) "Predicting Response to Hurricane Warnings: A Reanalysis of Data From Four Studies." Mass Emergencies 4:9-24.

Hanson, P.O., Vitek, J.D. and Hanson S. (1979) "Awareness of Tornadoes: Importance of an Historic Event." Journal of Geography, January, 22-25.

Hodler, T.W. (1982) "Residents' Preparedness and Response to the Kalamazoo Tornado." Disasters, 6(1):44-49.

Glass, R.I., et al. (1980) "Injuries From the Wichita Falls Tornado: Implications for Prevention." Science, 207, 734-738.

Sims, John H. and Baumann, Duane D. (1972) "The Tornado Threat: Coping Styles of the North and South." Science, 176, 1386-1392.

Book references

Leik, Robert K., Carter, T. Michael and Clark, John P. (1981) Community Response to Natural Hazard Warnings. Minneapolis, MN: University of Minnesota.

Steele, Alex, Lyons, Morgan and Smith, Donald D. (1979) Disaster Contingency Planning: Area Agency on Aging. Tallahassee, FL: Florida Research Center.

Taylor, J.B., Zurcher, L.A. and Key, W. H. (1970) Tornado: A Community Responds to Disaster. Seattle: University of Washington Press.

Technical papers

Baker, Earl J., Brigham, John C., Paredes, Anthony J. and Smith, Donald D. (1977) The Social Impact of Hurricane Eloise on Panama City, Florida. Florida Sea Grant Technical Paper, Gainesville, FL: University of Florida.

Baker, Earl J. (1984a) "Response to Warnings and Probabilities in Alicia." Paper presented at the National Hurricane Conference, April 26-28, Tampa, FL.

Baker, Earl J. (1988) Hurricane Evacuation Behavior in the Middle Atlantic and Northeast States. Tallahassee, FL: Hazards Management Group, Inc.

Technical reports

Baker, Earl J. (1980) Public Response to Hurricane Frederic: Variation Within the Warning Area. Environmental Hazards Center Report, Tallahassee, Florida: Florida State University.

Baker, Earl J. (1981) Evacuation of Galveston in Hurricane Allen. Environmental Hazards Center Report, Tallahassee, Florida: Florida State University.

Baker, Earl J. (1984b) Public Response to Hurricane Probability Forecasts. NOAA Technical Memorandum NWS FCST 29, Silver Spring, MD: National Weather Service.

Baker, Earl J. (1985) Eastern North Carolina Hurricane Evacuation Study. Tallahassee, FL: Hazards Management Group.

Baker, Earl J. (1986) Warning and Evacuation in Hurricanes Elena and Kate in Panama City, Florida and Pinellas County, Florida. Environmental Hazards Center Report, Tallahassee, FL: Florida State University.

Baker, Earl J. (1990) "Evacuation Decision Making and Public Response in Hurricane Hugo in South Carolina." Natural Hazards Research and Applications Information Center Quick Response Study Report 39, Boulder, CO: University of Colorado.

Bates, Frederick L., Fogleman, Charles W., Parenton, Vernon J., Pittman, Robert H. and Tracey, George S. (1963) The Social and Psychological Consequences of Natural Disaster. National Research Council Disaster Study 18. Washington, DC: National Academy of Sciences.

Killian, Lewis M. (1954) Evacuation of Panama City Before Hurricane Florence. National Research Council Disaster Study, Washington, DC: National Academy of Sciences.

Moore, Harry E., Bates, Frederick L., Layman, Marvin L. and Parenton, Vernon J. (1963) Before the Wind: A Study of Response to Hurricane Carla, National Research Council Disaster Study 19, Washington, DC: National Academy of Sciences.

Treasure Coast Regional Planning Council. (1983) Treasure Coast Hurricane Evacuation Study. Stuart, FL: Treasure Coast Regional Planning Council.

Wilkinson, Kenneth P. and Ross, Peggy J. (1970) Citizens' Response to Warnings of Hurricane Camile. Social Science Research Center Report 35, State College, MS: Mississippi State University.

Windham, Gerald O., Posey, Ellen I. and Spencer, Barbara G. (1977) Reactions to Storm Threat During Hurricane Eloise. Social Science Research Center Report 51, State College, MS: Mississippi State University.

THE ROLE OF TECHNICAL HAZARD AND FORECAST INFORMATION IN PREPAREDNESS FOR AND RESPONSE TO THE HURRICANE HAZARD IN THE UNITED STATES

T. MICHAEL CARTER
Hazards Management Group
1146 19th Street, NW
Suite 500
Washington, DC 20036
U.S.A.

After the model has been calibrated for a given area, hundreds of hypothetical hurricanes are "run through" the model to determine each storm surge inundation pattern. The hypothetical hurricanes are systematically chosen to reflect variations in intensity, point of landfall, angle of approach and forward speed. Each model run presents the storm surge height for a number of model grid points, both along the open coast and well inland.

Because of the uncertainty inherent in forecasting the exact point of landfall of any actual hurricane, the results of the individual model runs are aggregated to identify what may be called the "storm surge floodplain." For example, all individual model runs of storms with maximum winds of 130 miles per hour, making perpendicular landfall in the model area, and with forward speeds of 10 miles per hour are aggregated to give the maximum surge height found for each individual model grid point. The results of these aggregations identify the coastal areas subject to storm surge inundation for particular hurricane scenarios and, thus, the areas that must evacuate during an actual threat situation resembling the particular scenario.

The vulnerability analysis

After the individual model runs are aggregated to form a number of hurricane scenarios, the patterns of storm surge inundation are overlaid on detailed maps of the area. Threatened populated areas and critical facilities are determined. A series of analyses are then conducted on the threatened areas. First, demographic data are compiled for the threatened areas to determine the number of people to evacuate. Second, an analysis of the roadway system is conducted to determine the evacuation routes to use and their traffic capacities. Finally, critical facilities--e.g., hospitals, nursing homes, water works, etc.--within the inundated area are identified for special planning attention.

J. Nemec et al. (eds.), Prediction and Perception of Natural Hazards, 75–81.

Based primarily on the population distribution and the roadway system, the threatened area is then divided into discrete "evacuation zones." These zones become the focus of planning efforts by local officials.

The behavioral analysis

Social science research has shown that, to be effective, evacuation plans developed by emergency management agencies must be based on realistic assumptions of how the public will behave when advised or ordered to evacuate. For this reason, detailed surveys of public response to hurricane evacuations have been conducted for most storms that have affected the U.S. coast over the past two decades. Taken together, these surveys provide comprehensive data on patterns of public response in a variety of hurricane threat scenarios.

In particular, these surveys provide data on what percentage of the population in different risk areas evacuate, when they evacuate relative to the evacuation notice, how many vehicles they utilize in the evacuation, where their destinations are and what type of facility they utilize as shelter. Each of these variables is analyzed with respect to the nature of the hurricane scenario--i.e., primarily the intensity of the threatening hurricane. These data are utilized to develop realistic assumptions concerning the probable behavior of the public in response to each of the hurricane scenarios developed during the vulnerability analysis.

Evacuation time estimates

The final step in developing the Technical Data Report is to compute estimates of the amount of time that will be required to complete evacuations of the threatened areas for each of the identified hurricane scenarios. These estimates are based upon quantitative transportation models that utilize the demographic data for each evacuation zone, the roadway system capacity analysis and the behavioral assumptions. These analyses are usually regional in focus, since evacuees from neighboring jurisdictions will typically use common transportation corridors after leaving their neighborhoods. These analyses also focus on identifying individual intersections or roadway segments that are subject to serious traffic congestions during evacuations.

Utility of the technical data report

When completed, the Technical Data Report gives local emergency management agencies information t use in developing detailed hurricane evacuation plans for their communities. In particular, for a variety of hurricane scenarios, the Report identifies which areas to evacuate, how long it will take to complete the evacuation, what portions of the transportation network will require traffic control measures and what proportion of the evacuees will utilize public access shelters.

However, to effectively utilize this information during an actual hurricane threat situation, the local officials must be able to accurately determine which of the hurricane scenarios presented in the Technical Data Report they are facing. It is at this point that the use of real-time hurricane forecasts becomes critical.

The hurricane forecast program

In the U.S., "official" hurricane forecasts are prepared and disseminated by the National Weather Service's National Hurricane Center, located in Coral Gables, Florida. This center is staffed by a group of experienced tropical meteorologists whose efforts are devoted exclusively to hurricane forecasting and research.

From its birth as a tropical depression until its dissipation, each tropical cyclone in the Atlantic basin is the subject of a continuing series of forecasts produced by the National Hurricane Center. These forecasts are produced on a regular schedule--every six hours--and are disseminated to federal, state and local officials, as well as other interested parties in the private sector.

The forecasts contain estimates of a number of relevant tropical cyclone parameters for the current conditions and 12, 24, 36, 48 and 72 hours into the future. The forecast parameters include the cyclone's position (the latitude and longitude of the storm's center), the speed of the storm's maximum sustained winds and the radius of winds of 34, 50, 64 and 100 knots, as appropriate.

Responding to the hurricane threat

Theoretically at least, local officials are able to plot each successive hurricane forecast on a hurricane tracking map to determine the extent to which their community is threatened. They then determine which pre-identified hurricane scenario the storm most closely resembles. Having identified the appropriate hurricane scenario, the officials can use the information in the Technical Data Report to determine the required response actions--which areas to evacuate in what time frame.

The theoretical part of this use, of course, involves forecast error. While the National Hurricane Center's official forecast is undoubtedly the best information available, it will be wrong on one or more key parameters. To date, there has not been a "perfect" hurricane forecast, and the hurricane forecasters assure us there will not be one in the foreseeable future. As a result, local officials and other users of hurricane forecast information must "bound" the forecast parameters with estimates of how large the forecast error may be.

Utilizing hurricane forecast error

Since the National Hurricane Center both estimates the current conditions of each storm and forecasts it's future conditions every six hours, it is in a position to collect and analyze very detailed forecast error statistics. Utilizing these data, we can compute the amount of "anticipated" forecast error for a variety of probability levels.

In terms of making response decisions, there are three primary forecast parameters which that be considered: where the storm will go (location of the track), when it will get there (timing along the track) and how intense it will be. Thus, after plotting the official hurricane forecast on a map, local officials must place "confidence intervals" around the forecast to account for the anticipated forecast error. Based upon these confidence intervals, local officials can then determine the "worst probable scenario" that they face and can use this scenario as the basis of their response actions. This worst probable scenario has high credibility since it is based explicitly on the real-time official forecast in combination with historically observed forecast errors.

Estimating hurricane forecast error

A variety of techniques have been developed to estimate the confidence intervals that should be placed around the official forecast parameters. Along with the official forecast, the National Hurricane Center issues "Hurricane Strike Probabilities" for all major cities along the coast. These probabilities, which are based on the historical forecast error data, represent the likelihood that the center of the storm will pass within about 75 miles of each city, given the specific forecast track. These probabilities are further given for a number of future time intervals so that they contain error information on both where the storm might be expected to go (location) and when it might be expected to get there (timing). The probabilities do not, however, give forecast error information on the storm's intensity.

Another technique, has been developed by the Hazards Management Group, a private consulting firm. This technique involves a graphic computer program that plots the official forecast positions and overlays on these positions the forecast error ellipses for user selected probability levels. The position forecast errors are represented as ellipses because the "along track" (timing) errors are larger than the "across track" (location) errors. The resulting graphic is a conical confidence interval around the official forecast track that is superimposed on a map of the coast. In addition, this program presents intensity forecast error information in terms of the probability that the storm's intensity will exceed that which is forecast.

Utilization of the program by local officials

As one would expect, not all local officials fully utilize the full range of available information to reach their hurricane evacuation decisions. The reasons are numerous and illustrate many of the problems involved in attempting to communicate technical information on the nature of natural hazards to the practitioner community.

First, the impact of a hurricane on any given community is a very rare event. On the average, the "return" interval of a hurricane to a community is longer than an official's term of office. Thus, most hurricane landfalls represent the responding officials' first hurricane experience.

In such a situation, comprehensive training programs must be used to replace actual experience. While sufficient funds were committed to conduct the research to produce the Technical Data Reports, no funds were committed for a continuing program to train local officials in the use of these data. The programs that do exist consist primarily of tabletop drills or exercises that are conducted locally by the individual jurisdictions or regionally by state agencies.

Given the limited knowledge of most local officials of the more technical aspects of hurricane forecasting techniques, the forecast errors inherent in these techniques and the implications of these errors for appropriate evacuation decision making, most local officials simply rely on the "advice" of local weather service offices. That is, they typically initiate hurricane evacuation when told to by local weather service personnel. However, since most local weather service personnel have very little tropical meteorology background, they rely, in turn, on the generalized advice of the National Hurricane Center.

In addition to the official forecast, the National Hurricane Center issues Public Advisories that describe in general terms the storm's status and the nature of the threat it poses. Within the context of this public warning program, the National Hurricane Center issues a Hurricane Watch for a portion of the coast when the hurricane is estimated to be about 36 hours from landfall. A Hurricane Warning is issued for a smaller portion of the coast when the hurricane is estimated to be about 24 hours from landfall.

To account for the anticipated error in forecasting the point of hurricane landfall, the average 36 hour position forecast error (about 200 miles) is added on each side of the forecast landfall position to determine the Hurricane Watch area. Likewise, the average 24 hour position forecast error (about 125 miles) is added on each side of the forecast landfall position to determine the Hurricane Warning area.

Traditionally, that is prior to the introduction of the comprehensive hurricane preparedness and response program described in the earlier sections of this paper, evacuations were ordered only after the Hurricane Warning was issued. The Hurricane Warning area adequately takes into account where the storm might hit (location) but does not take into account the forecast errors in when the storm might hit (timing) or how intense it might be when it hits. In fact, it was a recognition of these shortcomings in the utility of the Hurricane Warning that led to the initiation of the Hurricane Preparedness and Response Program, the dissemination of the official forecast to local officials and the introduction of the hurricane strike probabilities.

The issuance of the Hurricane Warning, in fact, forms the primary basis of the evacuation advice given to local officials by local weather service personnel.

Evaluation of the U.S. hurricane preparedness and response program

Like many government sponsored hazard-related programs, the Hurricane Preparedness and Response Program has lost something in its implementation. Without an identified and continuing "technology transfer" function, technical information provided to the practitioner community merely accumulates dust on a library shelf.

Relevance for other natural hazards

The hurricane is probably the ideal meteorological phenomenon on which to base a comprehensive preparedness and response program of the type described here. It is a well organized weather system that is readily distinguishable from its environment. As a result, its forecast parameters are easily separated from each other and easily verified.

This is apparently not the case with synoptic systems that cause heavily rain or snow events or outbreaks of severe local storms. Thus, no useful forecast error statistics have been compiled by research meteorologists that could be applied to the three relevant response questions of where, when or how intense. While the author is convinced, as a nonmeteorologist, that the compilation of such statistics are possible, such an undertaking may well be an exercise in futility in the absence of a focused technology transfer function.

The primary issue that must be addressed in programs dealing with various aspects of preparing for and responding to natural hazards involves how to bridge the chasm that separates the technical researcher--who focuses on the hazard on a daily basis--from the practitioner--who does so only occasionally.

REFERENCES

Journal articles

Neumann, C. J. and Pelissier, J. M. (1981) "An Analysis of Atlantic Tropical Cyclone Forecast Errors." Monthly Weather Review, 109: 1248-1266.

Technical reports

Baker, E. J. (1983) "Public Response to Hurricane Probability Forecasts." National Weather Service, Silver Spring, MD.

Carter, T. Michael (1983) "Probability of Hurricane/Tropical Storm Conditions: A User's Guide for Local Decision Makers." National Weather Service, Silver Spring, MD.

Federal Emergency Management Agency (1984). "A Guide to Hurricane Preparedness Planning for State and Local Officials." CPG 2-16, December, Washington, DC.

Jelesnienski, C. P., Chen, J. and Shaffer, W. A. (1992) "SLOSH: Sea, Lake and Overland Surges from Hurricanes." National Oceanic and Atmospheric Administration Technical Report, Silver Spring, MD.

U.S. Army Corps of Engineers (1984) "Technical Guidelines for Hurricane Evacuation Studies." November, Washington, DC.

U.S. Army Corps of Engineers (1987) "Eastern North Carolina Hurricane Evacuation Study." Technical Data Report, November, Wilmington, NC.

U.S. Army Corps of Engineers (1987) "Eastern North Carolina Hurricane Evacuation Study. Appendix B: Hazards Analysis." Technical Data Report, November, Wilmington, NC.

U.S. Army Corps of Engineers (1987) "Eastern North Carolina Hurricane Evacuation Study. Appendix C: Behavioral Analysis." Technical Data Report, November, Wilmington, NC.

SESSION 3

UNCERTAINTIES IN THE DEVELOPMENT OF PREDICTIONS OF FLOODS AND LANDSLIDES

SOME REMARKS ON SCIENTIFIC ACTIVITY OF FLOOD PREDICTION IN ITALY

LUCIO UBERTINI
Università di Perugia - Ist. di Idraulica Agraria
Borgo XX Giugno, 74
06100 - Perugia
Italy

INTRODUCTION

Non structural actions for flood control have had a very consistent development in the last twenty years for many reasons and in particular:

1) the reliability of real time hydrological monitoring systems;
2) the accuracy of flow and flood forecast by simple mathematical methods;
3) the high cost and the high environment impact of the structural works for flood control.

In 1984 the Italian Ministry of Civil Protection in cooperation with the Ministry for Scientific Research and the Ministry for Public Works established the National Research Group for Prevention of Hydrogeological Hazards. The Group is incorporated within the National Research Council and it must develop methods and technologies for the control of the hydro-geological hazards taking into consideration similar achievements at the international level.

The policy of the Group involves the coordination of projects, scientific and normative activities, including aspects of institutional, local administrations, legislative bodies and private companies and professional enterprises in order to attain optimal use of resources for research actions and special programmes.

During the last five years the Group has been very active in the area of methodologies that could be of use in flood hazard prevention and mitigation. In particular, significant results have been obtained with regard to real time flood forecasting models, and floods originating from breaking of dams or levées. In the following sections this paper discusses some of the results obtained in these two areas.

J. Nemec et al. (eds.), Prediction and Perception of Natural Hazards, 85–92.

Real time flood forecasting models

In the scientific literature, the possibility to select simple adaptive models for real time flood forecasting has been widely discussed. These models may be appropriate to forecast lead times which are short in comparison to the response time of the basin.

Two such simple models, the RS model (an extension of the unit hydrograph adaptive model to cover large basins) and the CM model (which has a relatively more complex structure but does not increase the computing task significantly) have been proposed by Corradini et al. (1986, 1987). These models provided reasonable flood predictions for a basin with mean lag time of 15-18hr and forecast lead times of up to 6hrs. Both models consider variability in space of rainfall and losses and incorporate the updating of flow forecasts at each time step substantially through the correction of one parameter involved in the description of losses.

The Upper Tiber River basin with a drainage area of 4147 km^2 was used for analyzing the performance of the models. The basin, located in the Umbria Region (Central Italy) is characterized by a complex orography. The topography is mainly hilly with mountain peaks making up a large portion of the basin boundary. The soil is generally made up of clay and sandy silt and the underlying bed is practically impervious rock. Floods are normally caused by widespread rainfall in the autumn-winter period. The annual average precipitation over the basin, mostly in the form of rain, ranges from 700 to 1600 mm.

Thirteen rainfall-runoff events with hourly data were utilized in this study. Five of these events were used for estimating the parameters to determine in advance. The estimated parameter values were: time of concentration 17.6 h, storage coefficient 11.2 h, and saturated hydraulic conductivity O.4 mm h^{-1}. Eight isolated events were employed for testing the models. These events, observed in the period 18 December 1981 through 24 March 1983 (Corradini and Melone, 1986), were characterized by peak flows ranging from 323 to 843 m^3s^{-1} and baseflow ranging from 39 to 135 m^3 s^{-1}.

Applications of these models to the events of the prediction set were carried out in conjunction with those of the unit hydrograph adaptive model (Chander and Shanker, 1984) whose initial delay time was estimated to be 7 h (Corradini et al., 1986). The behaviour of each model in relation to each other was practically similar for forecast lead times of up to 6 hours.

Specifically, a reduced increase in model result accuracy was found using the CM model instead of the RS model, with average values in peak flow error (in magnitude) of 9.3% and 12.9%, respectively. However, the unit hydrograph method produced very large errors with an average value of 63.5%. These results were also confirmed by an analysis of the shape characteristics of the curves predicted. An indication about the adequacy of the basic structure of a given model may be derived from the behaviour of the adaptive parameter for which, time-invariant values in a given event should be expected if the model is perfectly suited. In reality, the CM model considers in the adaptive scheme a combination of the two parameters, ie. the sorptivity (S) and a coefficient (C), however, useful information of the model structure may be obtained by considering only the variability of the parameter S provided it is estimated at each time of flow forecast (Corradini and Melone, 1986). The variability of each parameter, for each event, was analyzed through the width of its range of values referred to the mean range value. It was observed that the variability of the infiltration parameter, I for the RS model, and φ, for the unit hydrograph based model, are similar, while the variability of S is not directly comparable with that of these two parameters. Thus, in each event, the variability in infiltration depth, $\Omega(S)$, produced in each homogeneous zone by the variability of S was computed. As a representative example, we provide for this quantity the results obtained in the largest homogeneous zone. For the eight events of the prediction set, the following extreme values were computed for the ranges of I, φ and $\Omega(S)$: 10-95%, 5-90% and 15-45%, respectively. Corresponding average values of about 55%, 50% and 30% were obtained.

Flood prediction from dam or levee failure

Submersion waves originating from the breaking of a dam and of a levée differ in two main characteristics: the spatial display of the wave and the value of potential energy generating the flow field. In fact the dam-breaking wave usually flows in a narrow valley river bed and it is essentially a unidimensional phenomenon. Levée breaking produces two-dimensional flows in the close ground plane, or a channel network flow if the direction of the flow is constrained by fixed obstacles as in the case of an urban road network flooding.

As regards the hydraulic head, values of the order of 100m are not unusual for the reservoir head behind a dam, while values of the order of a few metres are usually reached by the water level of a levée-constrained river. As a consequence, different orders of magnitude characterize the potential energy involved in the two cases. Obviously, many different situations with scales contained between the two limiting cases quoted above-can take place, such as when a monodimensional dam break wave outflows a narrow valley into a wide plane or when a break is produced in a long earth dam.

Several well known historical disasters have depicted the effect of dam breaking submersion waves on the environment and on the basis of such observations and collected data the causes, water levels and volumes, velocities and time scales have been evaluated.

The water outflowing from a total or partial break in a dam originates a wave characterized by a peak value and a time to peak determined by the time of breach formation. The analysis of several cases has shown that even though it takes time for the formation of the breach, it may be of very short duration, and hence "for the sake of safety", the breaking is often schematized as instantaneous.

The mass of water falling into the valley river bed from a height, of maybe 100m, forms a wave advancing downstream, characterized by an abrupt front and followed by a more gradual varying wave-body. Due to the very high bottom slopes typical of mountain valley river beds, supercritical flow conditions occurring at the dam location are usually maintained further downstream. Transitions from supercritical to subcritical flow conditions and vice versa may easily occur due to the effect of the often irregular geometry of the valley cross-section, and increased resistance to the bed modification produced by the travel of the wave.

Usually the simulation of the dam-break submersion wave is required in order to evaluate some significant characteristics of the wave, useful to foresee the impact on the environment. The maximum stages reached by the water along the entire valley, the time required by the front to reach particular points where bridges, towns, and other structures are located, the speed of the water, are correlated with impact effects such as, erosion, transportation, etc.

The simulation of the whole phenomenon is naturally very difficult and, of course, requires several arbitrary hypotheses in order to quantify essential characteristics which are impossible to foresee. The results of the simulation must be used to evaluate the order of magnitude of the involved variables and their sensitivity to the governing physical effects. Arbitrary and sometimes unjustified hypotheses are usually introduced to simulate the following aspects:

1) **Breach Formation:** The modality of breach formation is quite different for concrete dams and earth dams (Johnson and Illes, 1976; McDonald and Landridge-Monopolis, 1984). In the first case, more than the collapse of the structure itself, the collapse of the rock holding its foundations and landslide waves overlapping the top of the dam are cause of the disaster. Both such geotechnical variables are difficult to predict and quantify. The most common causes of earth dam breaking are connected with water overtopping and foundation soil fluidification piping. The dimension and time required for the breach formation have to be assigned on the basis of reasonable hypotheses or even of unreasonable extreme ones like the abrupt dam collapse. A spectrum of simulations based on different excitations will be obtained as a result.

2) **Water-bed interaction:** It is practically impossible to foresee the modification of a valley produced by the propagation of the dam break wave and, as is obvious, this effect is a determinant for the evolution of the phenomenon itself. Material is abruptly taken from the valley sides and transported downstream, structures met by the wave front can be destroyed, transported and subsequently accumulated in narrowings of the valley. The obstructions so originated can produce subsequent backwater effects and further collapse.

3) **Submersion wave propagation:** The problem of simulating the routing of the wave, even through an undeformable bed, can be approached either from physical or numerical basis. Physical simulation is often preferred in the first stages of the phenomenon and when the valley is characterized by abrupt changes of direction, depth and section form, so that form effects, hardly accounted for by a mathematical description, can be properly taken into account.

Following the distinction introduced above, we shall deal separately with the two situations of plane submersion wave produced by levée breaking and urban area flooding. In both cases the flow field is spatially two dimensional, even if the flooding of an urban road network can be schematized "naturally" as an equivalent network of channels connecting accumulation points.

Referring to the plane submersion wave, we can distinguish, quite naturally, the part close to the breach, the "near field", where the flow is dominated by inertial effects and pressure forces and the "far field" where the flow is influenced essentially by the ground characteristics: slopes, roughness, saturation, etc. The near field flow is rapidly varied and characterized by a small space scale while the far field flow can reasonably be schematized as gradually varied and displaying over a large scale field.

As regards the "near field', the physical simulation is more suitable than the numerical one especially when dealing with complex geometries and the impact of the submersion wave against obstacles. Both the scale of the problem and the dominating forces are in this case, favourable for the physical simulation. When geometrical boundaries are not too complex, in this case too useful information can be obtained from the numerical simulation. The flow is mathematically described by the well known shallow water equations written in divergence form: the flow variable is either the unit discharge vector or the mean velocity, the geometrical variable is the water height. The correct definition of the boundary conditions according to the subcritical/supercritical regime flow is slightly more complicated than in the one dimensional case and more difficult problems are met in finding the numerical integration schemes well suited for the regime. Many questions are still not well defined as classified even at physical level, especially as regards the front propagation over dry bed, and the two-dimensional abrupt front propagation. In fact, even if computer techniques are still at research level, promising results have already been obtained with simplified field geometries (Vila, 1987; Katopodes, 1984b; Van, 1981). Also in this case, the numerical problem is considerably simplified neglecting inertia terms in the shallow water equations for higher water level simulations but considerable qualitative discrepancies are obtained in the "near field".

As to the large-scale problem, the characteristic time and modalities of the levée breach formation are not such important features as for a dam; moreover, inertial effects can safely be neglected in a simulation aimed at obtaining an estimate of breaking and maximum water levels to produce inundation maps.

Due to these simplifications and to the scale problem, the numerical simulation is in this case, better suited than the physical one (Cunge, 1975; Laura and Wang, 1984). Practical problems are nevertheless connected with the evaluation and correct definition of the plane roughness, that often should have local directional properties and should be assigned as a second order tensor. Laboratory experience has pointed out the importance of the soil saturation in determining the wetted front advancement (Van, 1981).

As for the numerical simulation of an urban area flooding caused by the levée breaking of a river or levée overflow, the main direction of flow can be fixed a priori on the basis of a simplified urban road network (Braschi et al., 1988). The storage effect can be usefully assigned to the nodes of the network itself wile assigning a kind of "urban porosity" to the topological area pertaining to the node. The road slopes are often much steeper than channels, the roughness not well defined, and hence form resistance and local losses due to fixed and moving obstacles are quite important and almost unpredictable. The channel flow equations are simplified neglecting inertial effects but upstream or downstream dependence of the flow must carefully be taken into account to get the correct solution for a complex network.

Conclusion

In accordance with the resolutions of US-ASIA conference on 'Engineering for the mitigation of natural hazards damage' (Bangkok, 14-18 December 1987) and of the International Workshop on Natural Disasters in European-Mediterranean countries (Perugia, 27 June to 1 July 1988), the priority of research activity must be assigned to flood prediction.

The results of the real time flood forecast models can be improved with a large research effort mainly in these following directions:

1) a better understanding of the role of infiltration at the basin scale;

2) the spatial-time rainfall distribution measured and processed in real time to furnish, possibly, the rainfall nowcast which must be introduced in the model.

The research activity on hydrological problems of dam and levée rupture is to improve existing procedures and to develop new procedures for estimating the flood and sediment hydrograph resulting from the failure of dams and levées. The specific objectives are:

1) To identify and test an appropriate model to determine the progression of the dam or levée breach resulting from different triggering mechanisms, overtopping and sudden collapse of part of the structure. Various boundary conditions must be considered as

well as different designs of the structures.

2) To develop and test mathematical and physical models for estimating the flood hydrographs from dam and levée breaches. A number of factors need to be investigated more accurately. These factors include: the geometry of the embankments, structural materials, initial breach configuration, boundary conditions, effect of sediment and flow conditions and geometry of the channel.

It should be stressed, however, that the scientific community, though conscious of its own limits has been, is, and will continue to be available to furnish its own contribution, because of its appreciation that work on these problems is very satisfactory not only from the scientific, but also from the social point of view.

References

BRASCHI G. and GALLATI M. (1987) Simulation of flooding in a two dimensional field. Proceedings of 18th Modeling and Simulation Conference, Pittsburg.

BRASCHI G., GALLATI M. and NATALE L. (1988) La simulazione delle inondazioni in ambiente urbano. Int. Rept. no 152 of Dept. of Hydraulic and Environmental Engineering, University of Pavia, Italy.

CHANDER S. and SHANKER N. (1984) Unit hydrograph based forecast model. Hydrol. Sci. J. 29, 279-291.

CHERVET A. and DALLERES P. (1970) Calcul de l'onde de submersion consecutive à la rupture d'un barrage. Schweiz Banz, 88, no. 19.

CORRADINI C. and MELONE F. (1986) An adaptive model for on-line flood predictions using a piecewise uniformity framework. J. Hydrol., 88, 365-382.

CORRADINI C., MELONE F. and UBERTINI L. (1986) A semi-distributed adaptive model for real-time flood forecasting. Water Resour. Bull., 22 (6), 1031-1038.

CUNGE A. (1970) Calcul de propagation des ondes de rupture de barrage. La Houille Blanch, 1, 15-23.

CUNGE J.A. (1975) Two-dimensional modelling of flood plains. Unsteady Flow in Open Channels, Water Res. Publications, Fort Collins.

CUNGE J.A., HOLLY F.M. and VERWEY A. (1980) Practical aspects of computational river hydraulics. Pitman Advanced Publishing Company, London.

GALLATI M. and UBERTINI L. (1989) Hydrological problems of dam and levée rupture. Technical Conf. on the Hydrology of Disasters, Geneve, 2-3 Nov. 1988, WMO.

JOHNSON F.A. and ILLES P. (1976) A classification of dam failures. Water Power Dam Constr., 28, 43-45.

KATOPODES N.D. (1984a) A dissipative Galerkin scheme for open channel flows. Proc. ASCE J. Hydr. Eng., 110, no. 4.

KATOPODES N.D. (1984b) Two dimensional surges and shocks in open channels. Proc. ASCE J. Hydr. Eng., 110, no. 6.

KATOPODES N.D. and CHIEN-TAI WO (1986) Explicit computation of discontinuous channel flows. Proc. ASCE J. Hydr. Eng., 112, no.6.

LAURA R.A. and WANG J.D. (1984) Two dimensional flood routing on steep slopes. Proc. ASCE J. Hydr. Eng., 110, no. 8.

LAURENSON E.M. (1964) A catchment storage model for runoff routing. J. Hydrol., 2, 141-163.

LINSELY R.K., KOHLER, M.A. and PAULHUS J.L.H. (1982) Hydrology for Engineers. McGraw Hill, New York.

Mc DONALD T.C. and LANGRIDGE-MONOPOLIS J. (1984) Breaching characteristics of dam failures. Proc. ASCE J. Hydr. Eng., 110, 567-586.

REED D.W. (1984) A review of British flood forecasting practice. Tech. Report no. 90, Institute of Hydrology, Wallingford, UK.

SMITH R.E. and HEBBERT R.H.B. (1979) A Monte Carlo analysis of the hydrologic effects of spatial variability of infiltration. Water Resour. Res., 15(2), 419-429.

SMITH R.E. and PARLANGE J.Y. (1978) A parameter-efficient hydrologic infiltration model. Water Resour. Res., 14(3), 533-538

VAN M. (1981) Modèle bidimensionnelle d'onde de rupture de barrage. Comparaison mesures-calcul pour un cas schematique, E43/81-08. Laboratoire National d'Hydraulique, EDS, France.

VILA J.P. (1987) Schemas nuperiques en hydraulique des écoulements avec discontinuités. Proc. IAHR, Lausanne, 120-125.

LANDSLIDES AS POLYCAUSAL PHENOMENA

DAVID E. ALEXANDER
Department of Geology and Geography
University of Massachusetts
Amherst, MA 01002
U.S.A.

According to Burton, Kates and White, (1978) the impact of natural hazards is a function of the benefits associated with living in a risk zone minus the costs of damage and of adjustment to the hazard. Alternatively, UNDRO (1982) argued that the sum total of risk is a product of the magnitude and extensiveness of the natural hazard impact, the number and size of elements at risk, and their vulnerability in terms of probable levels of damage and destruction.

In certain fields the impact of natural hazards is fairly immutable, or subject to definable natural trends. In others it can be altered by human activity, which affects the number of elements at risk. This vulnerability is a function of the amplification or mitigation of risk (Alexander 1991). One precondition for either of these opposing tendencies is risk perception. Risk perception may either increase or reduce human vulnerability and conditions a range of actions, from planning decisions to public behavior. It represents the prevailing consensus on how much importance should be attached to the recognition and management of natural hazard risk.

Such schemes work well for earthquakes, but are less easy to apply to landslides, which are essentially *polycausal* phenomena. As the causes of slope instability may include both natural and anthropogenic factors (not to mention natural factors that are exacerbated by human activity), the categories of impact, risk and vulnerability are more difficult to separate from one another. In this paper I will discuss three cases of slope instability that result from diverse combinations of factors, in which natural causes cannot necessarily be separated from anthropogenic ones.

Debris slides and avalanches at Cuyocuyo, Southern Peru

Cuyocuyo is situated in the Cordillera Oriental of southern Peru, about 225 km north of Lake Titicaca. In its vicinity, Ordovician, Silurian and Devonian slates have been

J. Nemec et al. (eds.), Prediction and Perception of Natural Hazards, 93–99.

compressed into an anticlinorium, with open and chevron folds and many overthrusts. The Rio Cuyocuyo rises on the Carabaya Altiplano at altitudes of 4500-4650 m and flows northeastwards along a valley that is about 1000 m deep (from the floor to its rock shoulders) towards the Amazon basin. The valley sides are mantled with a talus of rock fragments that forms cones and aprons (derived respectively from point and areal sources) which stand at angles of 32°-42°.

Sideslopes have been extensively terraced for somewhere between 550 and 1500 years, with the downslope faces of each terrace composed of an often elaborate stonework. Although many of these structures have been carefully maintained, in such a high energy environment they are not sufficient to counteract all erosional processes and landslides.

Several recent mass movements had serious consequences. The road north from Cuyocuyo passes close to the eastern valley side where it undercut the toe of the slope. Shallow talus slips have occurred, covering areas of up to three and a half ha. Cones formed where debris is canalized by structural weaknesses. These can be very unstable if undercut. One destroyed a group of houses in the mid-1980s and the occupants narrowly escaped with their lives.

Coarse clastic debris has rolled from eroded sideslopes into the Rio Jilari, a tributary of the Rio Cuyocuyo. Severe floods in 1983 and 1984 developed a flow almost two km long in the saturated rock waste. The flow entered Cuyocuyo town, creating a large debris fan, destroying a church and many houses, and killing two residents. According to local witnesses, the movement was a cross between a debris flow and a hyperconcentrated streamflow (cf. Wieczorek 1986), but with very coarse, slaty debris in it.

At Nacoreque, approximately three to four kilometers north of Cuyocuyo, two major debris avalanches occurred during prehistoric times. The Nacoreque II avalanche involved about 600,000 m^3 of rock and sediment, which slid on a 33° slope and blocked the Rio Cuyocuyo to a depth of 100 m. Stratigraphic evidence suggests that: a) the avalanches occurred before the onset of the Holocene, when weathering conditions differed from current ones, b) that these movements were of the extremely rapid kind (*sturzstroms*--Hsü 1975), (c) they both impounded landslide-dammed lakes; and d) the lakes progressively breached in a series of major floods (cf. Costa and Schuster 1988).

In all the Cuyocuyo landslides, natural causes predominate over anthropogenic ones. Six episodes of orogeny affected the central Andes from the Late Paleozoic to the Pliocene epoch. Collectively, these have created a wide variety of faults, joints and folds that are still amenable to selective erosion, including landsliding and debris avalanching. At least four glaciations have occurred during the Quaternary. The landscape is responding by sporadic and belated adjustments to a series of events that have disturbed its equilibrium. Debris production and the energy of relief have both been very high, and landslides are an inevitable consequence of this. Human intervention, such as the oversteepening of slopes, has had relatively little impact on slope stability, but human vulnerability in such an environment remains persistently high.

Debris flows at Calciano in Basilicata, Southern Italy

The small municipality of Calciano is situated at 700 m above sea level on the southern flank of the River Basento. It is located in the Basilicata Region, 62 km from the Ionian Sea in southern Italy. The immediate vicinity of the settlement consists of seven contiguous drainage basins with a total area of 1058 ha and a northward slope that descends towards the Basento Valley at 10°. The local area is dominated by alternating bands of Miocene sandstone and grey marly clays. A few patches of clayey sands with lenses of cobbles and gravel rest on top. Four debris flows and two gully complexes occur in the main deposits.

The locality has long been forested. A map made in 1910 shows the hillsides completely covered by undisturbed forest. Subsequent deforestation and the degradation of woodlands has led to slope instability. In the 1960s 77 ha of upland at altitudes of above 400 m were gradually cleared of trees. Thus 18.9 percent of the land area of two central catchments was converted into arable fields. Runoff and seepage from the fields augmented springs at an altitude of about 500 meters and initiated a debris flow-slide. The slide eventually overwhelmed three concrete check dams and now threatens the Naples-Taranto railway, which passes along the Basento Valley close to the basin outflow point.

At Calciano, anthropogenic causes predominated over all others. While tectonic and lithostratigraphic conditions undoubtably conspired to make mass movement likely, the principal factor was deforestation and consequent alteration of local groundwater hydrology. Runoff and throughflow from newly established wheat fields fed a pre-existing line of seepage at the contact between outcropping marls and limestones. The pore water pressures were sufficient to mobilize more than 125,000 m^3 of debris in the channel and on its sideslopes. Thus, groundwater seepage was favored by contacts between exposed limestone and shale strata. The pattern of these outcrops provided structural control of the evolving landslide. Soil interflow seeping out into gullies sapped their heads and turned them into active mudflow sources.

Deforestation at Calciano can be seen in terms of a general tendency during the present century to eradicate forests in Basilicata and to increase the area of pasture and arable cultivation.

Debris flows in the Valle dell'Orco, Basilicata, Southern Italy

The Torrente Orco and its neighbor, the Torrente Sativo, drain northward into the River Basento 84 kilometers from the Ionian Sea coast near Vaglio Basilicata. The Orco-Sativo basin (referred to here as the Valle dell'Orco) is 675 ha in size, and is situated at elevations of 580-1115 m above sea level. It has an overall slope of 8°. The northerly 10 percent of the basin consists of allochthonous Miocene clays, silt-clays, limestones and jaspers. The remainder is composed of variegated clays, marls, calcarenites and sandstones of the Oligocene Corleto Perticara formation. Both of these are highly erodible. The basin currently has a mean annual rainfall of about 800 mm, with the marked seasonality that is characteristic of a Mediterranean climate.

From 1955 to 1980 the Valle dell'Orco underwent a dramatic transformation. An extensive but relatively superficial gully system metamorphosised into a complex of mudflows and debris flow-slides that eventually occupied 271 ha or 40 percent of the basin. Inundation of the drainage system with mudflow material reduced the total length of channels in the basin from 56.9 km in 1955 to only 19.96 km in 1980. Drainage density was reduced from 8.43 to 2.96 km/km^2.

The most obvious potential cause of this extraordinary transformation is the process of rationalization and mechanization of agriculture. The period from the 1950s to the 1980s was one of profound changes in the rural economy of Basilicata. In the 1970s agricultural occupations for the first time accounted for less than 30 percent of the regional labor force, while unemployment rose to 15 percent and employment in the tertiary sector rose by 82 percent. Natural population growth fell from 10.7 percent in 1951 to 1.5 percent in 1981 and the resident population shrank by 13 percent, a result of the drift to the northern cities and European industrial heartland. The population of the highlands of Basilicata was smaller in 1981 than in 1881.

Reduction in the pressure of population on the land permitted consolidation of holdings into bigger units while mechanization enabled the cultivation of larger, more regularly shaped fields. The introduction of fertilizers led to the abandonment of fallowing in favor of annual wheat crops with no rotation. Eventually, European Community farm subsidies made it economically feasible to abandon the most unstable and least productive land regardless of ownership. As a result of these changes, the number of farms in the Valle dell'Orco catchments fell from 25 in 1955 to 13 in 1980. The field pattern was drastically altered by the consolidation of holdings and mechanization of tilling and contained significantly fewer boundaries. The total area of managed land decreased by 117.5 ha or 17.4 percent, in proportion to the abandonment of arable fields.

In the Neogene clay terrains of Basilicata, mass movements, or landslides, are strongly associated with gully incision. They tend to occur near gully headcuts and meander bends and to be stimulated by ploughing. The ploughing concentrates rainwater infiltration at the bottom of the disturbed layer and creates prolonged throughflow. It is, therefore, reasonable to suggest that deeper ploughing and more intensive cultivation (without fallow) have been responsible for increasing the water and sediment yield into gullies and thus initiating the mudflows. (Alexander 1982) This hypothesis is confirmed by the location of large areas of present and past arable land at the headcuts of the mudflow complex that eventually developed. Land abandonment resulted, not merely in rural depopulation and agronomic change, but also in the increase in slope instability associated with evolving mudflow complexes. Reforestation was restricted to the flanks of the increasingly unstable gullies.

This explanation, however, is not a sufficient one. It ignores the natural causes of geomorphological change, such as active tectonics. To investigate the geomorphological effect of these, lineaments crossing the Valle dell'Orco catchment were mapped from aerial photographs dated 1955, 1974 and 1980. They are manifest as topographic alignments, changes in the textures and hues of soils as a result of differences in moisture content, and shadows cast by highly localized subsidence of the ground. These features probably represent the surface traces of fractures and faults. The lineaments show pronounced Apennine (NNW) and anti-Apennine (ENE) trends, with the remnants of an older (probably Mio-Pliocene) conjugate fracture system orientated NE and NW.

In the Valle dell'Orco in 1955, 15.03 km of channels were obviously aligned with the fracture lineaments. Many other segments of channel showed inflections where they crossed the fractures (it should be emphasized that the lineaments were not defined on the basis of alignment with channels, but with respect to more general topographic creases and other signs). By 1980 5.6 km of channels guided by lineaments were eliminated, but the proportion of channels following the fractures had risen from 26.4 percent to 47.2 percent. Thus, channels aligned with fault and fracture traces tended to be more resistant to obliteration by mudflows.

As a result, the main causes of the mudflow complex in the Valle dell'Orco can be summarized as: a) weak, disturbed geological formations with widely varying permeability and high plasticity, b) intense fracturing caused by neotectonics, c) drastic changes in land use and management; and d) the abandonment of agricultural land. Isolating any one of these causes is misguided. They are both complementary and mutually dependent.

Conclusion

The examples presented above illustrate three diverse sets of conditions. In the Peruvian case, slope instability is an inevitable natural phenomenon, as post-glacial and post-orogenic adjustment are both occurring in a high-energy geomorphic environment. At Calciano, the primary responsibility for the creation of a large and destructive mudflow rests with deforestation. Finally, in the Valle dell'Orco, persistent tectonic forces and changes in land-use are the dual culprits with respect to mudflow development on a large scale.

The common factor in all three cases is that human intervention (road construction at Cuyocuyo, deforestation at Calciano and agrarian change in the Valle dell'Orco) is capable of destroying fragile and temporary equilibria on slopes and in valleys. The degree of significance attributed to this depends upon whether natural causes are capable of accomplishing the same transition to instability, and whether they would do this on a time scale comparable to the anthropogenic one.

The examples in this paper represent cases in which human action caused the amplification of total risk. Risk mitigation measures would have been practicable, but given the inevitability of landsliding at each of the three sites, the most successful strategy would doubtless merely be to avoid amplifying the risk by unwise land-use. Only limited success could be achieved by more comprehensive (and expensive) strategies to reduce the natural causes of landsliding. This observation has important implications for the total costs of adjustment to slope failure hazards.

Another common factor in the three examples is that risk perception plays little or no role in reducing total vulnerability. Few attempts have been made to study attitudes to landslides among officials and the public (see Alexander 1989). However strongly the hazard was perceived, at the study sites cognition did not translate into an adequate strategy for avoiding or reducing the risk. For example, as Kayser (1964) pointed out, reforestation has met with limited success in reducing the slope instability that is widespread in southern Italy. In this case there has been a dichotomy between the perception of practitioners (farmers who ignored the problem) and that of the theorists (who perceived the problem, but mistook the solution).

Similarly, despite the presence on slopes of extensive pre-Columbian terraces, at Cuyocuyo landslides are simply accepted as one of the many inevitable dangers of life in the high Andes. Local resources of capital and manpower are not considered sufficient to be able to keep the problem under control. At Calciano and in the Valle dell'Orco, European Community subsidies (price support mechanisms resulting from the Common Agricultural Policy) tend to mitigate against soil conservation, as they provide no incentive even to consider appropriate measures. Hence, risk perception is biased by economic circumstances. These add a certain artificiality to risk management, as the reality of soil erosion is not addressed.

REFERENCES

Journal articles

Alexander, D.E. (1982) "Difference Between "*Calanchi*" and "*Biancane*" Badlands in Italy." In Badland Geomorphology and Piping, Bryan, R.B., and Yair, A. (eds), Norwich, U.K.: Geobooks.

Alexander, D.E. (1989) "Urban Landslides." Progress in Physical Geography.

Alexander, D.E. (1991) "Applied Geomorphology and the Impact of Natural Hazards on the Built Environment." Natural Hazards 4(1).

Costa, J.E., and Schuster, R.L. (1988) "The Formation and Failure of Natural Dams." Bulletin of the Geological Society of America.

Hsü, K.J. (1975) "Catastrophic Debris Streams (*sturzstroms*) Generated by Rockfalls." Bulletin of the Geological Society of America 86, 1975.

Wieczorek, G.F. (1986) "Debris Flows and Hyperconcentrated Streamflows." Proceedings of the ASCE Conference "Water Forum '86: World Water Issues in Evolution," Long Beach, California, 4-6 August.

Book references

Burton, I., Kates, R.W. and White, C. F. (1978) The Environment as Hazard. New York: Oxford University Press, New York, NY.

Kayser, B. (1964) Studi Sui Terreni e Sull'Erosione del Suolo in Lucania, Matera, Italy: Fratelli Montemurro Editori.

UNDRO. (1982) Natural Disasters and Vulnerability Analysis, Geneva: Office of the United Nations Disaster Relief Coordinator, Geneva.

UNCERTAINTY IN EVALUATING LANDSLIDE HAZARD AND RISK

ALBERTO CARRARA
National Research Council-CIOC
Viale Risogimento 2
Bologna
Italy

INTRODUCTION

In many countries landslides are widespread in space and time as to constitute the natural disaster which generates a yearly loss of property greater than that derived from any other catastrophe such as earthquakes, tornadoes and floods. This is largely true for Italy where the interaction of geological, geomorphological and climatic factors has led to a geomorphic evolution of the slopes mainly controlled by mass-movement. Hence, throughout the hilly and mountainous areas of the country, every year hundreds of failures take place damaging lifelines, dwellings and crop yields (Cotecchia, 1978; Canuti, 1982).

Different methods and techniques for evaluating landslide occurrence have been developed or proposed. Some of them aim at predicting the behavior of individual slopes through surface and subsurface measurements of soil/rock geotechnical properties and mechanical modelling of slope stability conditions. landslide incidence over large areas or whole countries on the basis of geological and geomorphological observations and models.

Despite the conflicting views between some earth scientists and engineers, it can be stated that both approaches have significant advantages and limitations (Cotecchia, 1978;Varnes et al., 1984, Hansen, 1984; Brabb, 1984; Nieto, 1989). Depending on the technical and economic constraints of the investigation, either the first or the second approach may be better suited to solve some slope instability problems or to fulfil the requirements of a specific project. In this short paper, discussion will be mainly confined to the analysis of uncertainty in predicting landslide *hazard* and *risk* on a regional scale, namely the assessment of actual and potential mass-movement over large areas derived from earth sciences information and modeling. Since a world-wide review of studies on this topic is beyond the scope of this work, the considerations presented in the following sections mainly reflect the author's experience gained from long-term investigations in Southern and Central Italy.

J. Nemec et al. (eds.), Prediction and Perception of Natural Hazards, 101–109.

Landslide hazard and risk zonation

Over the past twenty years, many scientists have attempted to assess *landslide hazard* and produce maps portraying its spatial distribution *(landslide-hazard zonation)*. The first term means "the probability of occurrence within a specified period of time and within a given area of a potentially damaging slope-failure" (Varnes et al., 1984). The second one refers to the division of the land surface into "homogeneous" areas or domains and their ranking according to degrees of actual/potential hazard due to mass-movement (Varnes et al., 1984). However, up till now there has been no general agreement on methods or even the scope of these in investigations (Brabb, 1984; Carrara, 1989).

Despite the methodological and operational differences, all methods proposed are founded upon a single basic *conceptual* model. This model requires first the identification and mapping of a set of geological-geomorphological factors (such as rock composition and structure, bedding attitude, slope steepness and morphology, stream evolution, vegetation cover, etc.) which are *directly* or *indirectly* correlated with slope-instability. It then involves both an estimate of the relative contribution of these factors in generating slope failure, and the classification of the land surface into zones of different hazard degree.

The model outlined - based on the well-known and widely applied principle *"the past and present are keys to the future"* - implies that slope failures in the future will be more likely to occur under those conditions which led to past and present instability (Varnes et al., 1984). Hence, the preliminary step of the operation implies the identification and mapping of all landslide phenomena over a *training area* which must be as representative as possible of the different geological and geomorphological features of the region to be investigated *(target area)*.

Experience gained from hundreds of surveys carried out in different parts of the world has demonstrated that well-trained investigators are able to detect and correctly map many or most of the landslides occurring in an area by applying aerial-photograph interpretation techniques and systematic field checks (Rib & Liang, 1978). However, old, dormant landslide bodies, landslide areas intensively modified by farming activity or covered by dense vegetation cannot readily be identified and correctly classified. This introduces a factor of uncertainty that cannot be readily evaluated and explicitly incorporated in the subsequent phases of the analysis, being largely dependent on the skill of the surveyor, and the quality and scale of aerial-photographs and base maps used.

As a point in case, it is worth noting that a detailed survey was recently carried out in a small drainage basin of Central Italy, the Tescio basin, where slope morphology is yearly modeled by intensive farming activity. Interpretation of high-quality aerial photographs, at the scale 1: 15,000, and systematic field observations permitted almost 250 landslides to be identified. For each failure the degree of certainty in its identification was estimated on the basis of the type and number of detected features which constitute the *geomorphic signature* of a failed slope (i.e. scarp lines, hummockey surfaces, undrained depressions, drainage lines pattern, etc.). Consequently, most landslides resulted classified with a high (63%) or fairly high (24%) certainty. The remaining ones (13%), consisting mainly of old, dormant, large slides, could be only be inferred (Carrara et al., 1990). Likewise, during a current study carried out in a basin of the San Francisco bay, California, three landslide maps were produced independently by different investigators. Their subsequent comparison revealed that the their overall agreement was fairly good, being of the order of 70%. However, several large, old, dormant/stabilized slides were interpreted by each surveyor in quite a different way (Carrara et al., 1992)

At present, many hundreds of maps of landslides or of their deposits have been produced throughout the world. With a few exceptions, their overall accuracy and reliability remain largely unevaluated, information on the criteria on which the failed masses were identified being insufficient or lacking. This notwithstanding, these inventory maps remain in general very useful for a preliminary assessment of the *actual* instability conditions of a region.

As previously outlined, a further step towards hazard zonation would require the identification of the conditions leading to slope-failure, their systematic and consistent mapping, and evaluation of their relative contributions to mass-movement in the area. All of this is not an easy task. In fact, the causes of each slope-failure are many, complex, and sometimes unknown. Conversely, the geological-geomorphological factors, which are both relevant to the prediction of landslide hazard that are *mappable* at an effective cost over a wide region are not many as yet (Carrara, 1989). Actually, the most important determinants of instability such as the hydrogeological conditions, are the most elusive to a quantitative measurement. Even worse, time and fund constraints frequently induce project managers and investigators to collect basic data from poorly reliable existing topographic and geological maps or obtain them by extrapolating a small number of site measurements of rock or soil properties to large areas. Hence the resulting information is largely inadequate for a reliable (as possible) prediction of hazard risk.

Other major problems refer to the methods commonly used in ranking slope-instability factors and in assigning the different hazard levels (cf. Brabb, 1984). This goal is accomplished through various procedures, among which the *geomorphological, index* and *statistical* ones are most important (Carrara, 1989).

All these methods incorporate significant advantages and drawbacks. The main advantages of the geomorphological approach lie in the capability of a skilful surveyor to estimate actual and potential slope failure, taking into consideration a large number of factors detected in the field or on the aerial-photographs. In addition, local or unique slope-instability conditions can be identified and assessed. The major drawback of the approach concerns the high *subjectivity* characterizing all phases of the geomorphological investigation. The degree of uncertainty associated to the different phases of the mapping operations cannot be evaluated. Likewise, it is difficult or sometimes impossible to compare landslide hazard maps produced by different surveyors. The outcome of the nation-wide project in France, "ZERMOS", (cf. Godefroy & Humbert, 1983) seems to point out clearly this limitation.

The greatest potential of the index approach consists in the *standardization* of the techniques for data acquisition, processing, analysis and display. This is true when it is accomplished with the aid of computer technology. In fact, Geographical Information Systems (GIS) permit the basic data to be readily acquired, manipulated and displayed. Pitfalls refer both to the large amount of time and money required to carry out studies over wide areas and, most importantly, to the *subjectivity* placed in rating and ranking factor contributions to landslide hazard. The resulting maps and information cannot be readily evaluated in terms of reliability or certainty.

The advantages and disadvantages of the statistical approach are similar to those of the previous method, with a *difference*: landslide hazard evaluation can become as objective an operation as possible, since the instability determinants are evaluated on the basis of a statistical model.

At present, a few investigators have attempted to predict landslide hazard through multivariate models or other geomathematical methods (Neuland, 1976; Reger, 1979; Carrara, 1983; Bernknopf et al., 1988; Carrara et al., 1990). This is quite surprising since the problem of forecasting landslides is conceptually and operationally similar to that of predicting other geological phenomena, such as oil traps or ore bodies, both of which have been extensively attempted by various geomathematical techniques (Harbaugh et al., 1977; Chung et al., 1988). In all cases the phenomenon to be identified is the result of the interplay of a large set of interrelated factors, many of which are unknown or unmappable.

The task of creating geological-geomorphological multivariate models enabling unstable slopes to be identified was attempted in three sample zones of Calabria (Southern Italy). The results demonstrated that these *black-box* models were capable of predicting actual and potential landslides in each study zone (Carrara, 1983, 1989).

More recently, by taking advantage of the potentials of modern GIS technologies, a new attempt was carried out in the previously mentioned Tescio basin. Through new techniques for automated cartography, a high-fidelity digital terrain model (DTM) was created and the study area was automatically partitioned into main-slope units (Carla' et al., 1987; Carrara et al., 1988). The resulting morphological parameters, merged with relevant lithological, structural and hydrogeological data, were used as input for a discriminant function which enabled the spatial distribution of basin stable and unstable slopes to be successfully predicted in over 80% of the cases examined (Carrara et al., 1990).

Although this multivariate approach proved to be more reliable than the commonly used empirical geomorphological techniques, it may not be readily applied for predicting any type of slope-movement. In addition, the model does not incorporate explicitly time, owing to the great difficulty in collecting adequate information on past occurrence of regional activation or reactivation of landslide bodies.

Once landslide hazard has been determined, the subsequent step should consist in determining the associated *risk* affecting property or lives. This task is neither conceptually nor operationally simple. In the literature risk related to mass-movement has been examined but rarely evaluated. The term itself seems not clearly defined; some authors use it to indicate the expected loss in lives and/or property due to slope failure (cf. Varnes et al., 1984; Brabb, 1984). Accordingly, landslide risk should be expressed by the product of landslide hazard and *vulnerability*. The latter is defined as the degree of loss to a given element (property/lives) resulting from the occurrence of a landslide of a given magnitude (cf. Brabb, 1984). Consequently, the assessment of the vulnerability of structures or human lives in relation to a certain slope-failure brings forth an additional factor of uncertainty. The problem becomes crucial when dealing with large, fast moving landslides such as rockslides or avalanches which may endanger urban areas or major engineering structures.

For the Tescio basin, the evaluation of risk (R_L) related to dwellings was attempted using a simple conservative approach. V_L was set equal to 1, which means total destruction of a building if affected by landsliding; therefore, risk was calculated as follows:

$$R_L = H_L \cdot A_L/A_s \cdot A_d/A_s \cdot D_L$$

where: H_L is the probability of occurrence of landslides within each slope-unit;
A_L/A_L is the ratio of unstable area to slope-unit area; that is, the expected size of unstable area within the slope-unit;
A_L/A_L is the ratio of inhabited area to slope-unit area; that is, the expected size of dwelling area within the slope-unit.
D_L is the degree of damage caused by landslide (ie. the expected loss of dwellings in a slope unit if affected by a landslide)

As previously mentioned, H_L was provided by the discriminant function. The discriminant model being time-invariant, H_L was tentatively assigned a time interval of 25 years, namely the return period of heavy rains in the region. By comparing the results of this operation (Fig. 1) with reports on the frequency of house damage caused by landsliding over the past 50 years, the simple approach selected appears to provide a fairly adequate model of the landslide risk existing throughout the study area.

Concluding remarks

The application of geomathematical methods to environmental hazards has always required a great deal of tine and personnel in collecting, encoding and digitizing the large amount of data needed to build-up multivariate models of the spatial distribution of landslides, floods and other natural catastrophes. To facilitate the acquisition, handling and

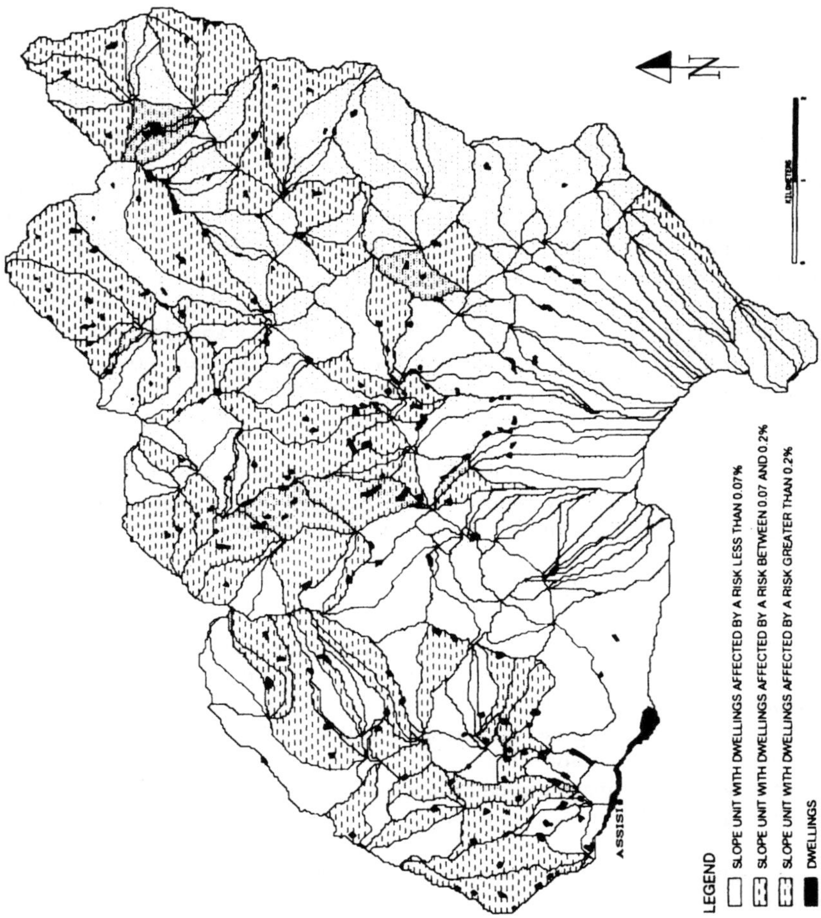

Fig. 1: Tescio basin (Central Italy) test area, landslide risk assessment

manipulation of these data, today efficient GIS-based techniques are available. The Tescio basin case study has clearly demonstrated the effectiveness of such an automated approach. Furthermore, sophisticated data capture/display electronic devices, faster computers and better GIS modules will profoundly influence the methods and techniques for acquiring and handling spatial data.

Discriminant analysis, based on a set of relevant geological-geomorphological variables mappable over a large area, proved to be able to predict stable and unstable slope-units in different sample areas. However, these statistical models do not lack limitations. First, they do not readily unravel the internal structure of the process involved. In addition, they remain basically too simple to fully explain slope-failure distribution, the latter being the result of the interaction of many factors, most of which are complex and some unknown. Likewise, extrapolating these functional models from training areas to large target areas may be difficult or sometimes impossible without an unacceptable increase in uncertainty. Hence, the resulting information cannot be used at the local scale without further detailed site investigations. In addition, in many cases the vulnerability of property or lives which could be affected by potential slope-failures cannot be readily determined, leading to a great uncertainty in risk assessment.

On the other hand, process-based models relying on systematic surface and subsurface geotechnical investigations over large areas, remain beyond any reasonable fund allocation. Moreover, at present stability analysis can be successfully applied to a limited number of slope-failure classes only (Nieto, 1989).

Consequently, as for other natural catastrophes, prediction of landslide hazard and risk remains a task largely unsolved both at regional and local scales. This would require a renewed interdisciplinary effort from research institutions, at the national and international level, supported by a financial commitment of the interested countries.

References

Bernknopf R.L., Carmpbell R.H., Brookshire D.S., Shapiro C.D., (1988). "A probabilistic approach to landslide hazard mapping in Cincinnati, Ohio, with applications for economic evaluation", Bull. Ass. Eng. Geol., v. 25, n. 1, p. 39-56.

Brabb E.E., (1984). "Innovative approaches to landslide hazard mapping", Proceed. IV Int. Symp. Landslides, Toronto, v. 1, p. 307-324.

Canuti P., (1982). "Ambienti geologici investigati nell'ambito del Sottoprogetto Fenomeni Franosi", Conv. Conclusivo P. F. Conser. Suolo, 9-10 giugno 1982, Roma, Sottopr. Fenomeni Franosi, Relaz. Gener., p. 227-234.

Carla' R., Carrara A., Detti R., Federici G., Pasqui V., (1987). "Geographical Information systems in the assessment of flood hazard", Int. Conf. on the Arno Project, Florence Nov. 24-24, 1986, p. 149-173.

Carrara A., (1983). "Multivariate models for landslide hazard evaluation", Mathematical Geol., v. 15, p.403-4Z6.

Carrara A., Cardinali M., Guzzetti F., (1992). Uncertainty in assessing landside hazard risk, ITC Journal, 2, pp. 172-183.

Carrara A., (1989). "Landslide hazard mapping by statistical methods: a "black- box" model approach", In: Siccardi F. & Bras R. L. (eds.), Inter. Workshop on Natural Disasters in European-Mediterranean Countries, Perugia, June 27-July 1, 1988, p. 205-224.

Carrara A., Cardinali M., Detti R., Guzzetti F., Pasqui V., Reichenbach P., (1990). "Geographical information systems and multivariate models in landslide hazard evaluation", Ninth Inter. Conference and Field Workshop on Landslides, Milano, Sept. 12, 1990, p. 17-28.

Carrara A., Detti R., Federici G., Pasqui V., (1988). "Reticoli idrografici e parametri morfologici di bacino da modelli digitali del terreno", Boll. Gruppo Naz. Difesa Catastrofi Idrogeol., Linea 3.

Chung C.F., Fabbri A.G., Sinding-Larsen R. (eds.), (1988). "Quantitative analysis of mineral and energy resources", NATO-ASI Series, v. 223, D. Reidel Pub. Co., Dordrecht.

Cotecchia V., (1978). Systematic reconnaissance mapping and registration of slope movements. Int. Ass. Eng. Geol. Bull., v. 17, p. 5-37.

Godefroy P., Hurmbert M., (1983). "La cartografie des risques naturels lies aux mouvements de terrain et aux seismes", Hydrogeol. Geol. Ingenieur, v. 2, p. 69-90.

Hansen A., (1984). "Landslide hazard analysis", In: Brmnsden D. & Prior D.B. (eds.). Slope instability, p. 523-602. Wiley, New York.

Harbaugh J.W., Dovedton J.H., Davis J.C., (1977). "Probability methods in oil exploration". Wiley, New York.

Neuland H., 1976. "A prediction model of landslips". CATENA, v. 3, p. 215-230.

Nieto A.S., (1989). "Mechanical models and geological observations: closing the prediction gap". In: Siccardi F. & Bras R. L. (eds.), Inter. Workshop on Natural Disasters in European-Mediterranean Countries, Perugia, June 27- July 1, 1988, p. 145-164.

Reger J.P., (1979). "Discriminant analysis as possible tool in landslide investigations",. Earth Surface Processes, v. 4, p. 267-273.

Rib H.T. & Liang T. , (1978). "Recognition and identification", In: Schuster R.L. & Krizek R. J. (eds.). Landslides analysis and control. National Academy of Science, Trans. Resear. Board Special Report, 176, p. 34-80, Washington D.C..

Varnes D.J. being (1984). "Landslide hazard zonation: a review of principles and practice" Commission on Landslides and other Mass-Movements, The Unesco Press, Paris.

THE RIVER DANUBE FLOOD OF 27 MARCH 1988*

ROBERT GEIPEL
Department of Geography
University of Technology, Munich
Germany

INTRODUCTION

European rivers tend to have long and well documented history of floods. The Danube, Europe's second largest river after the Volga, flows through or marks the border of eight states: Austria, Bulgaria, Czechoslovakia, Hungary, Romania, the Soviet Union, West Germany and Yugoslavia. As an international river, 2,857 km long, and with a catchment area of 817,000 km^2, it poses many ecological and political problems.

The German share of the Danube river basin amounts to 77,057 km^2, which is less than a tenth of the whole. This account is further restricted to a short reach of 176 km between Regensburg and Passau, the latter being the town where the Danube crosses the border between Germany and Austria, 2,202 km from its mouth and 280 m above sea level.

There is an abundance of research on flooding of the Danube. In the vicinity of Regensburg, the Schwabelweis gauge has measured river stage since 1845. The 20 largest floods to have occurred within the last 150 years have registered discharges between 1,910 m^3/sec. and 3,880 m^3/sec., while in unexceptional years the maximum flow is about 950 m^3/sec (fig.1). The flood of 27 March 1988 ranked fifth among the twenty largest events, with a discharge of 2,700 m^3/sec.

The left tributaries of the Danube originate from hill country and have their maximum flows in winter, while the right tributaries rise in the Alps and their maximum discharges occur in summer, at a time of melting snows. This means that the discharge of the main river is relatively balanced. The mean discharge of the Alpine River Inn exceeds that of the Danube itself, the tributary carries more water than the trunk stream it joins. Extreme floods result when maximum discharges from both halves of the river basin coincide.

* The author is grateful to David Alexander's linguistic improvements of the first draft of this paper

J. Nemec et al. (eds.), Prediction and Perception of Natural Hazards, 111–118.

Fig. 1: Probability of Floods at Schwabelweis gauge

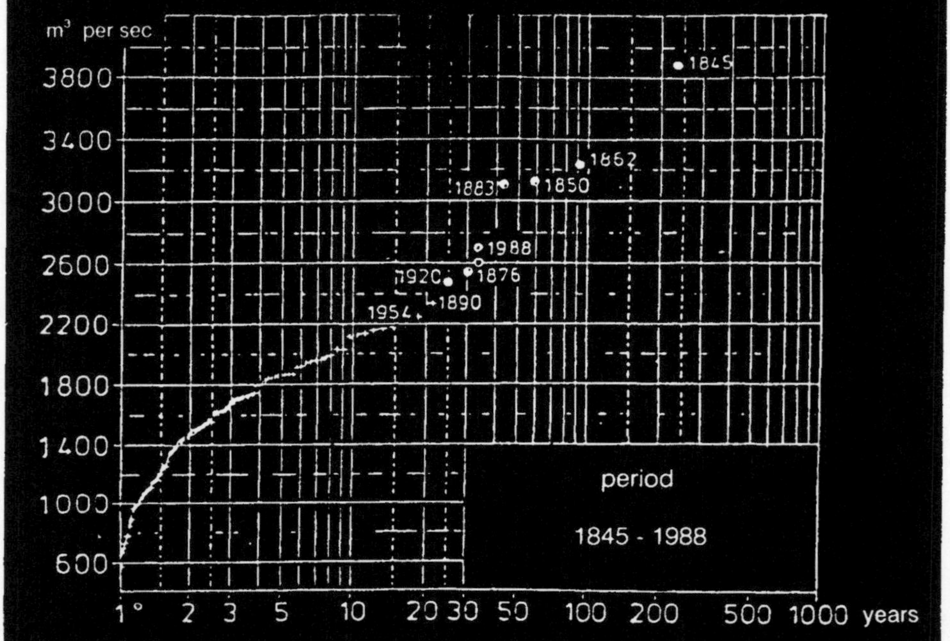

Source: Schiller 1988

Man-made changes in the extent of floods

The severity of flooding on the Danube has been strongly influenced by many changes occurring to its floodplain since the beginning of the 19th century. Moors and bogs have been drained, cultivated and settled, including the Danube moor, which measures 200 km^2. Since steamship navigation began in 1836, the meandering stream has been straightened. Artificial cut-offs through meander bends reduced the length of the Danube and caused it to incise its bed, thus increasing its fall and decreasing its width. The construction of levees began in 1884, with the aim of confining floodwaters to a narrower path.

In 1921 a contract for the construction of a ship canal was signed and the Danube was changed into a sequence of reservoirs with hydro-electric power stations. To stop the erosion of the bed, an agreement was signed in 1976 to change the whole of the Danube into a canal. These modifications led to a loss of floodwater retention space, as stagnant waters have been cut off and drained and the land filled. Riverine forests have been eliminated, wetlands drained and the infiltration capacity of soils diminished by field clearance.

The use of heavy agricultural machinery compresses soils and increases surface runoff. Crops like corn, sugar beet and turnips, which are planted in rows, increase the potential for soil erosion, especially during the spring floods where only scarce vegetation covers the fields. In recent decades the area planted with these crops has increased.

The construction of settlements and roads has sealed off large areas and increased the surface runoff. In the Alpine catchment areas of the Danube's tributaries leisure activities claimed vast areas for skiers, involving the clearance of forests and the construction of chair-lifts and downhill skiruns. Man-made increases of runoff added to the reduced storage capacity of forests which are dying under the impact of acid precipitation. The combination of all of these factors has increased the water yield of the catchment and speeded up the process of surface drainage, thus increasing the risk of flooding.

The Rhine-Main-Danube Canal

The section of the Danube most affected by the flood of March 27, 1988, is being rebuilt as part of the Rhine-Main-Danube Canal. The construction of this artificial water course has been harshly criticized because the estimates of commercial traffic that the canal is likely to carry are considered highly unreliable. Therefore, during the last years of the Social-Democratic government of West Germany, building was stopped, although it was resumed after the Conservative victory in the Spring of 1983.

The sectors where the levees were breached in 1988 contained hand-built dikes constructed between 1933 and 1942 to reclaim land for pasture and cultivation. These polders were drained by ditches and sluices. The dams and earthworks are now fifty years old. Local funds have not been spent on maintenance or repairs because the canal project is viewed as a major Federal expenditure on renovation of drainage and flood defenses.

But in the period of time that elapsed between the construction of weak gravel dams and work on the impending Rhine-Main-Danube Canal, the population of the floodplain has lived with increased flood risks. In the same degree that the upstream sector is well-proofed against flooding, the hazard in the downstream sector has grown. Present construction stopped before reaching Straubing (2,320 km from the mouth of the river), and it was here that the poorly-constructed levees burst. It can, therefore, be concluded that the history of the construction of the Canal will be closely linked with flood hazards on the Danube between Regensburg and Passau.

The controversy

Opinions on the Canal are divided between two highly confrontational parties. The first group considers this project as one of the most important works of the twentieth century, finally creating a connection between the North Sea and the Black Sea. This was a dream of Charlemagne, who in 793 A.D. failed in his attempt to build a small canal connecting

Rezat and Altmuhl via tributaries of the Main and the Danube. The second attempt involved the Ludwig-Danube-Main Canal, which opened in 1845 when the newly initiated railway age was bound to consign it to failure. The present project for a "Europe Canal," with a bank-to-bank width of 55 m, a bottom width of 31 m, a depth of 4 m and locks 190 m long, is being built to carry ships of up to 1,350 tons and is intended as a highly visible landmark of public power and political prestige (WIRTH 1986, p. 88). This view is shared by the renowned IFO Institute (Information and Forschung = research) (1982) and is also backed by many political statements.

The opponents of the scheme are environmentalists, critical economists (FULLING, 1978) and the German Council of Landscape Protection (1983) which--apart from the destruction of unique river valleys--lists the following as causes for rejecting the scheme:

(a) the expected decline in freight traffic (for example, the conversion of the Austrian steelworks Voest near Linz to new raw material);

(b) the excessive transportation time (from Rotterdam to Odessa by sea takes 6 - 10 days, while on the canal it will be 23 - 30 days);

(c) the threat of competition from East European countries, which may use the international canal to ship products at excessively low-priced freight rates;

(d) the ensuing decrease in freight traffic on the German Federal Railway, which already runs a high deficit;

(e) the plan to drain water from the River Danube into the dry Middle Franconian basin in a way that is less hygienic and economic than pipelines;

(f) the long-term damages involved in changing riverine meadows into arable land, thus endangering a unique riparian flora and fauna along the Danube downstream of Regensburg.

The people of the region, who were affected by the 1988 flood, are well aware of these inter-relationships.

The survey

We defined a study zone consisting of six communities on the north bank of the Danube and three on the south bank. In this area levees broke during the 1988 flood and 20 km^2 of land were inundated, necessitating the evacuation of people and livestock. We carried out questionnaire interviews with a sample of 180 households in the area. The rate of return, at an overall 57 percent (103 questionnaires), was higher on the north bank than on the south bank.

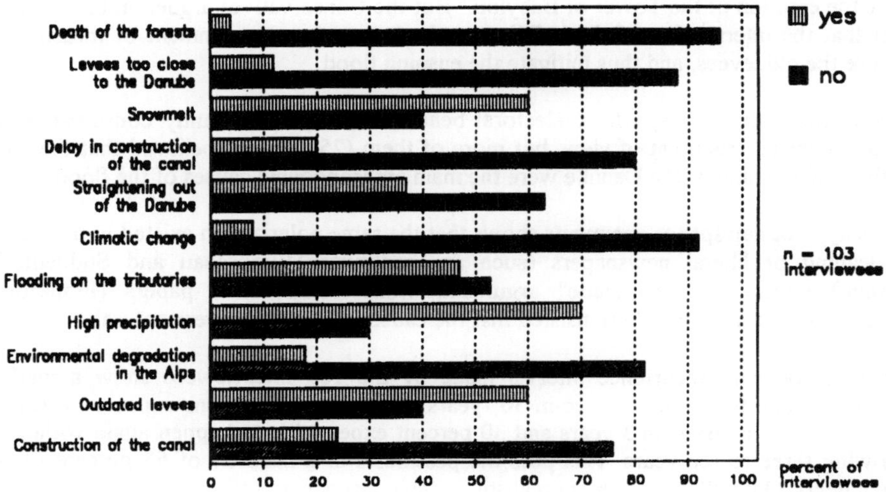

Fig. 2: Perceived causes of the flood of 1988.

The questionnaire, asked residents to assess the flood risk, define the likely recurrence interval for a flood of this size, give an explanation of the causes of the event, describe how they were affected by the disaster, explain their position or role in the local community and describe what responsibilities they had at the time. The 103 respondents who described what they perceived to be the cause of the March 1988 flood gave the following explanations (Fig. 2):

The natural cause was perceived as the coincidence of high precipitation (70 percent) and snowmelt (60 percent) in the mountains. Outdated levees, a man-made hazard, also attracted credence (60 percent) as a cause of the disaster, followed by floods on the northern tributaries and straightening of the course of the Danube. A minority of respondents regarded the construction of the canal (25 percent of interviewees) and the delay in building it (20 percent) as relevant causes of the flood. Environmental damage in the Alps and the death of forests rank last among the perceived causes.

The disaster has an indisputable political dimension, whether or not the culprit is regarded as construction of the Rhine-Main-Danube Canal or the temporary interruption of work during the Social Democratic-Liberal coalition in Bonn. Minister of Transport Hauff (Social Democratic Party) described the canal as "the most unnecessary building work since the erection of the Tower of Babylon." But the Conservatives argued in the Bavarian Diet that the interruption of the building works led to the abandonment of measures to restore the old levees, and thus mitigate the ensuing flood.

Among the interviewees, whose electoral behavior was predominantly conservative, 20 percent charred this point of view, but more of them (25 percent) believed that the canal or the straightening of the Danube were the main anthropogenic causes of the flood.

An analysis of newspaper comments shows that the same polarization existed in the media: left-leaning or liberal newspapers (such as Frankfurter Rundschau and Suddeutsche Zeitung) stress the "man-made" argument, while conservative papers (Frankfurter Allgemeine Zeitung, Die Welt) insisted that the causes of the flood were natural.

The predictions of recurrence interval made by the 103 interviewees show a marked pessimism. The flood was a once-in-30 -years event, but 25 percent expected a similar event to occur in less than 3 years and 30 percent expected it to happen again within the following three to ten years. This puts the pessimists in a majority of 55 percent of the respondents. Only 28 percent were realistic in their assessment (10 - 30 years) and a mere 17 percent were optimistic enough to expect a return period of 30 - 100 years.

Hardly any differences can be detected between the perceptions of north and south bank residents, even though the impact on the latter area was smaller. Evidently the population of the study area was sufficiently homogeneous for differences of perception to be slight. Respondents to a question about how to mitigate future floods put the responsibility firmly on the government.

Respondents tended to reject measures which would affect them personally, such as strengthening building codes, refusing people permission to live in the hazard zone or, even more drastically, relocating the local population. Technocratic activities were preferred to organizational ones. We are evidently dealing with a population that expects everything from the state which, since 1933, has improved living standards by making investments designed to increase the available arable land. At the same time it also endangered the population by not mitigating the flood risk, although promising permanent security in the near future with the completion of a magnificent construction project. Evidently the occupants of the flood-prone area along the River Danube in Lower Bavaria will have to repeat in the 1990s the experiences of the 1940s in the USA when Gilbert White started out to explore the then new field of natural hazards.

In line with the focus of our workshop, this paper is intended to show that the perception of flood hazards, and of scientific information regarding them, is filtered through the political opinion of those who are the target group of all predictions, forecasts and warnings--the population of the flood hazard area. Whether government policies appease this group or not depends strongly on the degree of partisanship between the local population and either the government or its political opposition.

REFERENCES

Bauer, R. (1965). Der Geschiebehaushalt der bayerischen Donau im Wandel wasserbaulicher MaBnahmen. In: Die Wasserwirtschaft, Heft 4 and 5, Stuttgart.

Broll, C.H. (1988) Der sinnlose Kanal. In: GEO 7, S, 98-117.

Harms, R.W. (1986) Auswirkungen der Urbanisierung auf den Hochwasserabfluß kleiner Einzugsgebiete. In: Schriftenreihe des Deutschen Verbandes fur Wasserwirtschaft und Kulturbau, H. 75, Hamburg, Berlin.

Kagerer, K. (1980) Donauausbau Regensburg - Vilshofen. Staustufe Straubing. Darstellung der Landschaftsentwicklung. Munchen.

Kasparson, R.E. (1969) Environmental stress and the municipal political system: the Brockton water crisis of 1961-1966. In: Kasperson and Minghi (Hrsg.): The structure of political geography. London, England.

Keller, R. (Hrsg.) (1979) Hydrologischer Atlas der Bundesrepublik Deutschland. Boppard.

Krankl, G. Historische Schutzmaßnahmen an der Donau zwischen Regensburg und Passau. In: Sonderdruck aus Informationsbericht Nr. 1/86 des Bayerischen Landesamtes fur Wasserwirtschaft.

Krankl, G. (1988) Hochwasserschutz an der Donau.

Landesamt Fur Wasserwirtschaft (Hrsg,) (1988) Endgultiger Bericht uber das Marzhochwasser 1988 der Donau, Munchen.

Orbig, K.H. (1988) Das katstrophale Donaùhochwasser 1988 zwischen Regensburg und Deggendorf. In: Wasserwirtschaft 79. Regensburg.

Popp, D. (1988) Hausgemachtes Hochwasser eroffnet Chance fur Neubeginn. Pressemitteilung des Bundes Naturschutz Sudbayern vom 14.4, Munchen.

Rhein-Main-Donau-Ag (1985) Rhein-Main-Donau-Wasserstraße, Munchen.

Rhein-Main-Donau-Ag (1988) Wasserstraße und Wasserkraftwerke, Munchen.

Schiller, H. (1985) Entwicklung und Ursachen der Hochwasserkatastrophen in jungster Zeit. In: Akademie fur Naturschutz und Landespflege (Hrsg.): Wasserbau-Entscheidung zwischen Natur und Korrektur. Laufen.

Schiller, H. (1988) Das Hochwasser im Deutschen Donaugebiet im Marz, Zwischenbericht.

Smith, K. und Tobin, G. (1979) Human Adjustment to the Flood Hazard, London, New York.

Unbehauen, W. (1973) Der Nachweis langfristiger Veranderungen der Hochwasseraß-flubverhaltnisse im bayerischen Donaugebiet. In: Deutsche Gewasserkundliche Mitteilungen, Sonderheft Munchen.

Unbehauen, W. (1971) Die Hochwasserabflußverhaltnisse der Bayerischen Donau. Hochwasser der Jahresreihe 1845/1965. In: Besondere Mitteilungen zum Deutschen Gewasserkundlichen Jahrbuch Nr. 30, Munchen.

Verfondern, M. (1982) Muß Passau mit dem Hochwasser leben? In: Der Erdkundeunterricht, H. 44, Stuttgart.

Whittow, J. (1979) Disasters. The Anatomy of Environmental Hazards. Athens.

Wirth, E. (1986) Die Ohnmacht der Vernunft. Vom Sinn und Unsinn eines Schiffahrtsweges Rhein-Main-Donau. In: Weiger, H. (Hrsg.): Der Rhein-Main-Donau-Kanal.

A SUMMARY OF THE STATE OF THE ART IN FLASH FLOOD WARNING SYSTEMS IN THE UNITED STATES

EVE GRUNTFEST
University of Colorado
Colorado Springs, CO 80933
USA

We are wrapping up a three year project assessing flash flood hazard in the United States. Our effort has three main components: 1) defining flash floods as opposed to slow rise floods; 2) reviewing the experiences of communities that are flash flood mitigation success stories; and 3) summarizing the advantages and disadvantages of warning systems. My presentation focuses on the third mission. Our work concentrates on flash floods in the United States but I am sure that some of our findings can be directly applied to international contexts and to landslide mitigation.

What we know about warning systems

Companies often sell these systems as "warning" systems. However, most of them attend to the detection part of the warning process and neglect the essential response components. Our findings are reviewed here in two parts: problems and pleasant surprises. This map (Figure 1) shows the location of the systems we studied. The paper concludes with research priorities and directions for the future.

Problems

Sweating the Small Things. Systems must be redundant. Any portion of the system that is not backed up is likely to fail. Mechanical parts are like to break down more frequently than expected. Batteries need replacement more often than vendors suggest. Emerson, Iowa reduced the number of transmissions to keep the batteries running more than two months. The vendors promised the batteries would last a year. Unanticipated equipment damage by weather, wildlife and vandalism occurs frequently.

Lack of Maintenance Funding. Warranties usually last one year. After that time communities, alone, must pay. The remote locations of many gauges present a major problem. The Ventura County Flood Control District estimates that maintenance of its

J. Nemec et al. (eds.), Prediction and Perception of Natural Hazards, 119–124.

system costs approximately $500 per gauge per year. A helicopter is required to change the batteries in the most remote gauges. Some communities chose the best textbook sites for gauges. While these are optimal data gathering sites, in some cases they are inaccessible. This increases personnel time and costs for maintaining the system. Heppner, Oregon's most remote gauge, is a full day's drive in a four wheel drive vehicle.

Wide Range of Levels of Protection and Expenditure. There is no consistent measure of how much money an effective warning system will cost. Costs for implementing the surveyed systems range from $500 to $4,000,000. There is no rule for what is an acceptable per capita expense for a warning system. The Lycoming County system which consists of a network of volunteer observers cost $500 to implement and serves a population of 118,000. At the other end of the spectrum is the Santee Cooper system. The Federal Energy Regulatory Commission insisted that the recently discovered earthquake hazard at Santee Cooper North Dam be addressed. Because of the small population at risk and the prohibitive cost of a new dam ($500,000,000) the commission approved implementation of a warning system and emergency action plan in lieu of structural modifications. The system consists of a seismograph and tail water detectors, which will alert dam operators if a breach occurs. Sirens, telephones and radios warn the downstream residents. The Santee Cooper warning system protects a population at risk of less than 700 residences at a cost of $4,000,000. The utility company is paying the entire bill, which compared with the costs of a new dam, is relatively cheap.

Inconsistent Level of Expectations and Formalization. Some systems require no training to be understood and operated, while others can only be managed by highly trained engineers. The software used in the Mount Airy system consists of five stages of alert that correspond to definite river stages. Although plans require the emergency management coordinator to be called in at the first-alert stage, actions are preset at each stage so anyone can give the order to evacuate. The software model used by the Santa Anna system is complicated any many assumptions built into the model such as the degree of saturation of the basin are understood only by a highly trained engineer.

Local and Individual Commitment to System. The systems that are most effective are ones in which officials have put a lot of energy into informing and involving the public. Systems resulting from federal initiatives are much less likely to be successful because local communities do not view them as their systems and are not committed to provide the equipment maintenance and replacement required to keep the systems operational. The Heppner system is a good example. The Corps of Engineers and the community decided to install a warning system. The Corps paid the upfront costs. The county was to assume responsibility for maintenance after the system was established. However, elections since the installation changed the county administration and the newly elected officials live far from Heppner. They have little interest in and commitment to the warning system. The Corps has assumed responsibility to keep the system running.

Charismatic individuals who take responsibility for the systems are crucial to success.

Kevin Stewart of Urban Drainage and Flood Control in Denver and Delores Taylor of Ventura County California have not only played key roles in keeping their systems going but in creating the team spirit within the ALERT user groups in the southwestern United States and in California.

Excessive Reliance on Technical Aspects/Underemphasis on Response Capability. Too many of the systems are simply detection systems. While the rationale for developing the systems is the reduction of loss of life and property damage, few of the systems pay much attention to getting the message to the populations-at-risk. Boulder has public education brochures and annual drills. However, the Pikes Peak region system for the metropolitan Colorado Springs, Colorado area has invested more than $250,000 in a system that will tell three men if they happen to be in their offices that there is less than one hour until people die. There is no mechanism, presently, to get the message to the people at-risk.

Warnings Are Only Part of a Comprehensive Flood Loss Reduction Program. There is tendency to see these systems as "paying off the raingods." Some communities find them an inexpensive alternative to politically unpalatable long term solutions that call for restricting floodplain uses and enforcing existing regulations. As Brian Hyde of the Colorado Water Conservation Board says, "development of a warning system is in some ways an admission of failure." If there were no communities in floodplains we would not need to develop detection and warning systems.

Pleasant surprises

Warning Systems as Data Collection Tools. The systems provide important data collection sources in areas where the hydrologic record is sparse. Forecasts are now based on a longer and more detailed record.

Peripheral Benefits. Maricopa County saves money by using its system to better estimate when public works officials will need to work overtime closing off flooded roads. The value of ALERT software and gauge networks have been used for lake quality monitoring, fire weather prediction, evapo-transpiration rates connected to urban area watering and toxic spill management. In California, the ALERT networks have been applied to drought mitigation in recent years when no floods have occurred.

In Aurora, Colorado, part of the Denver metropolitan area, water engineers rely on the ALERT software daily for water management decision-making. With the precise calculations of the evapotranspiration rates and the precipitation measurements that ALERT systems offer, the city requests releases for exactly what the requirements are. These reductions in wasted water for watering and for more precisely monitoring use rather than allotments saves the city vast sums of money annually, way in excess of the cost developing and maintaining the ALERT system.

Interagency Cooperation. The Pikes Peak region used the development of the system as an opportunity to initiate drainage wide discussions which make more sense than using the political boundaries. The Chesapeake and Ohio Canal National Historical Park and others required several local, state, and federal agencies to cooperate to get the system on line. An important side benefit of the systems is the interjurisdictional floodplain management they require.

Research Priorities

Briefly, we need ways to calculated the warning system benefits both tangibly and intangibly. Following Houston's 1988 flooding the public works department had no way of knowing how many lives were saved or how much property loss was prevented as a result of the warnings. The savings from the peripheral benefits could also be used to sell the systems to communities if we knew how to count them.

We need to have coordinated post-audits after floods to better understand which warning system components are working and which are not. We say we need floods to test out the systems but we should have a regular preflood-post flood study plan ready when the floods occur (Gruntfest, 1987).

We must address the false alarm issue directly. While I am convinced that false alarms occur infrequently and when they do people are still glad to have had a "dry run," many officials fear that one false alarm will undermine future public response. We have extensive experience with hurricanes in the U.S. where people who evacuated in response to warnings for storms that did not end up posing a major threat claim that they would respond to the next warning by evacuating as well.

We should package response systems parallel to detection system packaging. One reason there are more detection systems than response systems is that no vendors sell response systems. Each drainage system is different, but the vendors tailor their package to the local topography. No one follows these vendors around with the response end. Software packages are so easy to implement. Emergency managers are relying more and more on computer assistance. It is time for the response components to be incorporated in the initial phases of the adoption of detection/warning systems.

We need to calculate appropriate actions within specific lead times. This may be too difficult, but in order to get people out of their cars, we need to convince them they are better off wet than dead. In the United States, most flash flood victims lose their lives in their cars. Cars are perceived to be so safe. People must be better able to interpret the environmental cues and be willing to climb to safety outside their vehicles.

We need to be clear about where scientists disagree with regard to recurrence intervals. In the Rocky Mountain west there is great concern about the elevation at which flash flood producing rainfall will occur. Some scientists argue that the storms will not be very severe above 7500 feet. Others disagree and still encourage the placement of rain gauges above this elevation (Baker, et al., 1988; Jarrett, 1990; Mayer and Nash, 1987).

We must make sure that radio frequencies are maintained for the ALERT networks. Currently the federal government is challenging the flood warning frequencies with other possible uses.

We need a system of archiving and sharing the data for the various basins. Currently years of data sit in floppy disk boxes on various county officials' desks. These data are excellent sources for river basin research.

Finally, the interagency task force on warnings should develop warning system standards. As the Bureau of Reclamation and the Corps of Engineers choose warning systems, they must be more sensitive to local governments than they have been in the past or the projects are doomed. One community might be below dams owned by various private owners and more than one federal agency. The agencies must recognize that successfully developing warning systems involves a different set of skills and level of commitment than building a dam.

Directions for the future

Rich Getting Richer - Poor Getting Richer More Slowly. In the United States two trends increase the likelihood that large cities will have more means to reduce flood losses and smaller communities will have fewer resources: The Community Rating System and the Privatization of the Weather Service. Rather than go into details here, suffice to say that the small communities and rural areas are being left out of several service equations that can increase risk overall.

High Tech/High Touch. It is essential to realize that detection systems are not warning systems. All the components of the system must be operating if the objectives of the system are to be achieved. We must cultivate concern for both rare and frequent events. The Shadyside Ohio flood in June 1990 once again caught everyone off guard ("We didn't know this creek could turn into a raging torrent.") We will always be surprised by extreme events in small drainages. We will never have warning systems in all hazardous areas. We need to educate the publics about environmental cues more often than we do because in many cases they will be the only warning.

REFERENCES

Journal articles

Gruntfest, E. and Huber, C. J. "Toward a Comprehensive National Assessment of Flash Flooding in the United States." Episodes, v. 14, No. 1, 26-35.

Gruntfest, E. and Huber, C. J. (1989) "Status Report on Flood Warning systems in the United States." Environmental Management, v. 13, no. 3, 279-286.

Baker, V. Craig R. and Patton, P. (Editors) (1988) Flood Geomorphology, John Wiley, New York, 503 pages.

Gruntfest, E. (1987) Editor "What We Have Learned Since The Big Thompson Flood," proceedings of the tenth anniversary conference: Boulder, CO, University of Colorado, Institute of Behavioral Science, Natural Hazards Research and Applications Information Center Special Publication Number 16, 271 pages.

Jarrett, R. (1990) Paleohydrologic Techniques Used to Define the Spatial Occurrence of Floods in Geomorphology." v. 3: Amsterdam, Elsevier Science Publishers, 181-205.

Mayer, L. and Nash, D. (1987) Editors Catastrophic Flooding, Boston, Allen and Unwin, 381 pages.

SPATIAL PREDICTION OF LANDSLIDES

JORDI COROMINAS
Geotechnical Engineering Department
Civil Engineering School, Technical University of Catalonia
Barcelona
Spain

INTRODUCTION

A great diversity of methodologies relative to landslide prediction have been developed in recent years and especially, hazard and risk mapping have now become usual techniques. According to Hartlen and Viberg (1988) the complete landslide hazard evaluation should provide answers at least to the following questions: where will the landslides occur? which will be its volume and how far will it travel? when will it happen? what kind of movement will take place? how fast will it move? Even though all the questions are relevant, only the first two questions will be discussed in the paper.

Localization of unstable areas

The identification of potentially unstable areas is the first step in landslide prediction. Detailed analysis of a specific slope, considering geometry, groundwater pore pressures, soil or rocktesting and monitoring, are only undertaken in a restricted number of cases. When analyzing large areas more simple and cheaper techniques are needed. Methods commonly used vary from morphological analysis to data treatment techniques, as for example:

<u>Simple morphological indices:</u> Shallow slides and debris flows appear under specific topographical and lithological conditions. Concave slopes are more effective in collecting and concentrating subsurface water. Several indices relating catchment area characteristics with failure susceptibility have been proposed by Hatano and Okimura (cited by Oyagi 1984) and have the following form:

Hatano's index: $F = a^{0.33} \tan \theta$
Okimura's index: $F = A^{0.22} \tan \theta$

J. Nemec et al. (eds.), Prediction and Perception of Natural Hazards, 125–132.

where **A** is the catchment area for a selected point, **a** is the average length of the catchment area (obtained by dividing the area by the width of the foot line or the toe of a slide failure surface) and **θ**, the slope angle. In the Takedaira area of the Mizunami region, Hatano's critical value index for a slide zone generally exceeds 3.4 and Okimura's index at Tennodani in Rokko Mountain was found to be 2.96.

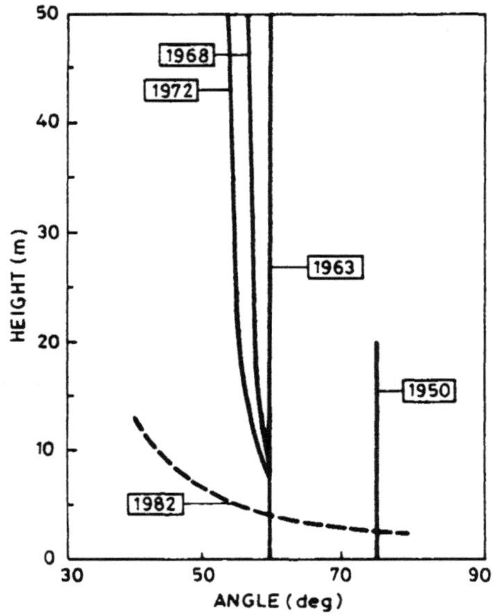

FIG. 1 Height versus slope angle plot for 177 failed and unfailed slopes in Hong Kong (modified from Brand and Hudson, 1982). Rules before 1982 have been discontinued. Dotted line represents a safety factor of 1.4.

The comparison with slopes of the same lithology and geographical region showing signs of instability has been traditionally used to establish threshold slope angles. Skempton (1952) found a critical slope angle of 10° in London clay natural slopes. Where groundwater level reached the surface in winter, slips can develop in 10° slopes, whereas flatter slopes were stable although subject to creep. This critical angle was in agreement with stability analysis based upon laboratory analysis of strength of soil samples. Nevertheless generalization based on this specific relationship is difficult due to the usual heterogeneous conditions of the slope. From an accurate analysis of 177 stable and failed slopes in Hong Kong, Brand and Hudson (1982) concluded that stability of slopes cannot be predicted on the basis of a few easily measurable parameters. Plotting height versus slope angle (figure 1), previous rules established for differentiating stable and unstable cut slopes were revealed inadequate. The dotted line corresponding to 1982 is the locus of slope geometries for which safety factor is 1.4 as calculated by conventional equilibrium analysis. This dotted line has a good acceptance as a lower bound envelope of the final geometry of failed slopes, although some slopes that have stood stable for more than 20 years fall above the line. This line then, should be considered as a long term stability slope behavior.

Data treatment techniques: Because slope stability depends on several factors acting at the same time, some efforts have been directed towards the acquisition of simply and quickly determined parameters. Stevenson (1977) using scored factors proposed a method to evaluate relative landslide risk in clayey slopes.

Discriminant analysis provides a more accurate stability assessment. A classical work using statistical techniques is that from Jones et al. (1961) on landslides in Pleistocene terrace deposits of Columbia River, bordering F.D. Roosevelt Lake. 160 slump-earthflow movement and additional 160 stable slopes were considered. Qualitative and quantitative factors influencing sliding were searched. A final analysis using the discriminant - function method was performed considering as influencing factors: original slope ($\mathbf{x_1}$) submergence percentage ($\mathbf{x_2}$), terrace height ($\mathbf{x_3}$) and groundwater ($\mathbf{x_4}$). The reduced discriminant function was given as follows:

$$y = 0.00216247 \log x_1 + 0.00334811 \log x_2 + 0.00944030 \log x_3 + 0.00673126\, x_4$$

Calculations of the discriminant functions for the 320 landslide and stable slopes gave values ranging from -0.0019 to +0.0404. Lower values represented stable slopes and higher values represented landslides. Less than one percent of the landslides had a discriminant value below 0.0106 which was taken as the limit of stability. Slopes with discriminant values between 0.0106 and 0.142 were considered as relatively stable because less than 5% of landslides fell between these values. A similar approach was used by Neuland (1976) who used the principal component analysis to select the influencing variables from an initial data set of 31 parameters. Final discriminant unction was obtained after rejecting some variables in order to avoid complexity. The discriminant function was:

$$y = 0.9222\times10^{-5}\, S^2 + 0.7926 \log (A + 10) - 0.6098 \log (d + 10)$$

where **S** is the slope gradient; **A** is the catchment area and **d**, a soil density related parameter. Values of 0.1934 mean unstable slopes and values of 0.1781 stable slopes. The model predicted correctly 94% of analyzed slopes.

Multivariate methods have been tested in Calabria, Southern Italy (Carrara 1983 a, b) to provide computerised hazard maps. Data relative to rock type, vegetation, slope gradient, surface roughness, stream frequency, presence of landslides among others had been stored into a computer-based data bank. The area was divided in cells of a 200m grid and a set of 14 geological and geomorphological variables were selected. After both discriminant and

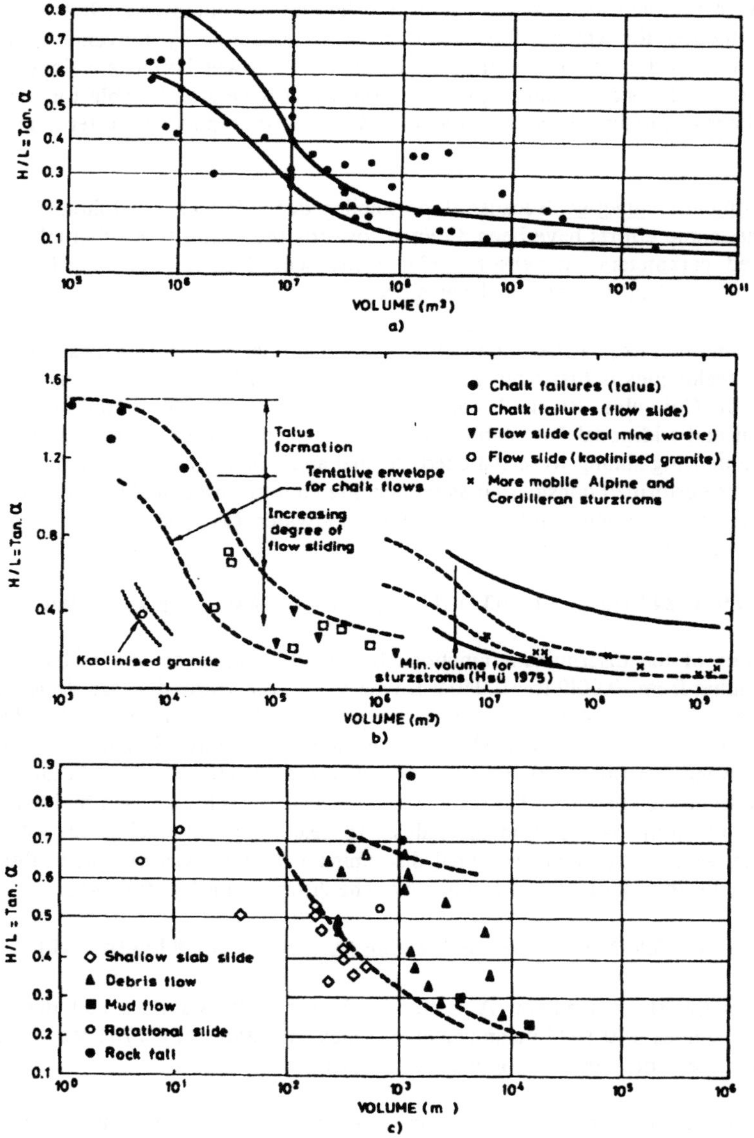

FIG.2 Landslide mobility. Volume of the fallen mass is plotted against height drop and horizontal distance travelled ratio. (a) data corresponding to huge rock avalanches (Hsü, 1975); (b) data for chalk, coalmine waste and kaolinized granite (Hutchinson, 1988); (c) data for small volume movements (Corominas et al. 1988).

multiple regression analyses, stable and unstable areas were mapped. Testing the accuracy of this method, give a percentage of cell between 73 to 83% correctly classified.

Landslide runout

The prediction of a hazardous zone involves not only the source of the expected landslide but the area that may be affected. Some huge rock avalanches reach long distances as for example, that from Nevado Huascaran in Peruvian Andes which travelled 16 Km demolishing the villages of Yungai and Rancahirca (Plafker and Eriksen 1979). Models taking into account the specific slope geometry, the strength of slided material and the slide mass path, as for example those for rockfalls (Lopez 1982, Rochet 1987) have usually been developed to explain the landslide runout. Nevertheless a relationship has been found between volume of fallen mass and reach. Hsü (1975) observed a decrease of the equivalent coefficient of friction (maximum drop height divided by the maximum horizontal distance travelled) with the rockfall volume (figure 2a). This friction decrease is more evident in biggest rockfalls. Hutchinson (1988) found a new envelope for chalk failures and flow slides of coal mine waste different from those proposed by Hsü. He observed a different reach for talus formation derived from chalk rockfall from that attained by chalk flows (figure 2b). Specific lithology and type of movement seem to have more important influence with volume reduction. Corominas et al. (1988) observed a different behaviour depending on the type of failure mechanism for less voluminous movements (figure 2c). This fact may be explained by considering that rock falls dissipate a lot of energy by hitting the substratum and breaking into small rock fragments, whereas shallow slab slides should only dissipate the energy as friction in the shear surface. Debris flows presenting a thick basal shear zone or dispersivity of their clastic components (grain flow), will occupy an intermediate place. The use of this empirical relationship is still limited due to the lack of more representative data.

For very small movements the use of the slope below the landslide source may have a significant influence on the progression of the displaced mass. This fact can be seen at the rockfalls inventoried in the Pyrenees area (Corominas et al. 1990) where the presence downslope of grassland, bedrock, scree deposits or forest is a conditioning factor for rockfall progression (figure 3). No rockfall paths have been found below the dashed line on the left. Above this line, the effect of the slope characteristics below the rockfall source is observable. Rockfalls propagate more easily in bedrock or grass covered slopes (tan α lesser than 0.75-0.8) than in scree and forested slopes. Central dashed line delimitate rockfalls paths through grass and bedrock slopes from scree slopes(tan α lesser than 0.8-0.85) and those ones from forested slopes. Given a rockfall source site and an expected volume to be mobilized and a specific topography, the reach of the rock fall mass may be determined as the intersection of the line dipping α degrees from de source with the ground surface.

As can be seen the volume is a conditioning factor of landslide mobility. Unfortunately its prediction is very difficult unless the failure has started and visible scars limit the landslide.

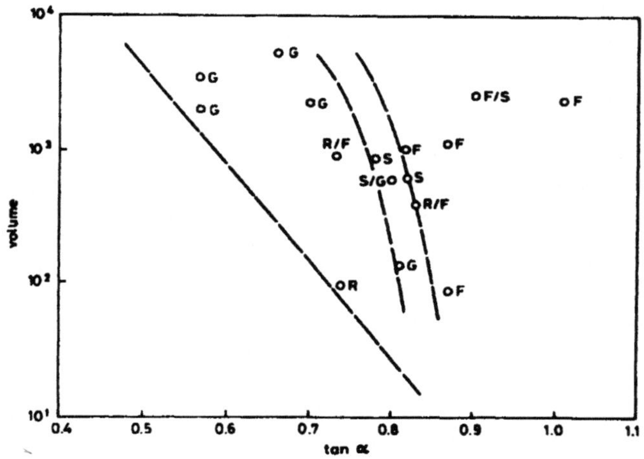

FIG. 3 Rockfall mobility. Plot of mobilized volume versus apparent coefficient of friction (tan α) derived from the path of the fallen debris, for several rock falls from Andorra and neighbouring Pyrenean area. Slope characteristics below the rock fall source: (G) grass (R) rock (S) scree deposits (F) Forest (Corominas et al. 1990).

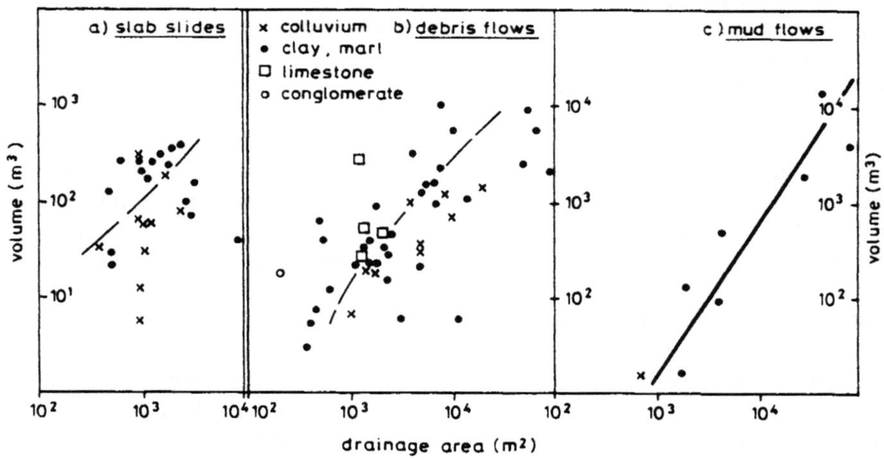

FIG.4 . Plot showing the relationship between watershed area and volume of the landslide for different kinds of mass movements induced by 1982 rains. Dashed lines in (a) and (b) provide a rough delimitation of failures in colluvium soils from other lithologies. A relatively good correlation is found for mud flows (c). (Corominas and Alonso 1990).

For shallow landslides Corominas and Alonso (1990) have found a rough relationship between the watershed area and the landslide volume (figure 4). The larger the watershed area the bigger the landslide in terms of mobilized mass. The fit seems to be better for mudflows whereas a lithological influence appears in debris flows and, to a lesser extent, in slab slides. landslide size, colluvium soils need to accumulate more water (larger watershed area) in order to fail than clayey substratum. This behaviour may be explained on the basis of better drainability of colluvial soils if compared with clayey soils. This relationship is not applicable to large landslides (greater than $10^6 m^3$) where, in movements inventoried in the Pyrenees area, an extra groundwater feeding coming from neighbouring watersheds or other factors seem to be necessary to trigger the slope failure.

Conclusions

The techniques that have been described in this paper represent a clear advancement in landslide prediction capability. Nevertheless most of these techniques are based on the assumption of a long-term invariability of soil and rock properties or on the maintenance of frictional parameters during the movement which often are not permanent. Uplifting in mountainous areas, cohesion decay by weathering or changing of physical properties in the surface of failure are responsible for some uncertainties. Some important questions as to which are the reconnaissance studies needed to be sure of the dimensions of a potentially landslide before the event , the loss of strength for large landslides or how to predict if an initial slide will transform itself into a mudflow or debris avalanche remain unanswered.

References

BRAND. E.W. & HUDSON. R.R. (1982). CHASE - an empirical approach to the design of cut slopes in Hong Kong soils. Proc. 7th South-east Asian Geotechnical Conf. 1: 1-16 Hong Kong.

CARRARA, A. (1983a). Multivariate models for landslide hazard evaluation. Math. Geology , 15: 403-426. New York.

CARRARA. A. (1983 b). Geomathematical assessment of regional landslide hazard. 4th Int. Conf. Applic. Stat. Probabil. in Soil and Estruct. Eng.: 3-27. Firenze.

COROMINAS, J., PENARANDA. R. & BAEZA. C. (1988). Identificacion de factores que condicionan la formacion de movimientos superficiales en los valles altos del Llobregat y Cardener. II Simp. Taludes y Laderas Inestables: 195-207. Andorra la Vella.

COROMINAS. J.& ALONSO, E. (1990). Geomorphological effects of extreme floods (November 1982) in Southern Pyrenees. Hydrology in Mountainous Regions II. IAHS. Publ. 194: 295-302.

COROMINAS. J., ESGLEAS, J. & BAEZA. C. (1990). Risk mapping in the Pyrenees area: A case study. Hydrology in Mountainous Regions II. IAHS Publ. 194: 425-428.

HARTLEN. J. and VIBERG. L. (1988). Evaluation of landslide hazard. In: Ch. BONNARD (Editor). 5th. Int. Congr. on Landslides, 2:1037-1057. Lausanne.

HSU. K.J. (1975). Catastrophic debris streams (sturzstroms) generated by rockfalls. Geol. Soc. Am. Bull.. 86: 129-140. Boulder.

HUTCHINSON, J.N. (1988). Morphological and geotechnical parameters of landslides in relation to geology and hydrogeology. In: Ch. BONNARD (Editor). 5th Int. Congr. on Landslides. 1: 3-35 . Lausanne.

JONES. F.O., EMBODY, D.R. & PETERSON. W.C. (1961). Landslides along the Columbia river valley, North-eastern Washington. U.S. Geol. Surv. Prof. Paper 367: 98p. Washington.

LOPEZ. C. (1982). Dinamica de los displazamientos de materiales sobre pendientes naturales. VII Simp. Nac. Obras Superf. en Mec. Rocas. 1: 2.5.1- 2.5.16. Madrid.

NEULAND. H. (1976). A prediction model for landslips. Catena. 3: 215-230. Cremlingen.

OYAGI. N. (1984). Landslides in weathered rocks and residual soils in Japan and surrounding areas: a state-of-the-art report. 4th Int. Symp. on Landslides, 3:1-31.Toronto.

PLAFKER. G. & ERIKSEN. G.E. (1979). Nevados Huascaran avalanches. Peru. In: B. VOIGHT (Editor): Rockslides and avalanches. Elsevier. 1: 277-314. Amsterdam.

ROCHET. L. (1987). Application des modeles numeriques de propagation a l'etude des eboulements rocheux. In: Risques Naturels. Bull. Liaison Ponts et Chaus.. 150-151: 84-95. Paris.

SKEMPTON, A.W. (1952). Stability of natural slopes in London Clay. Proc. 4th Int. Conf. Soil Mech.. 2: 378-381. London.

STEVENSON, P.C. (1977). An empirical method for the evaluation of relative landslip risk. Bull. Int. Assoc. Eng. Geol. 16: 69-72.

SESSION 4

UNCERTAINTIES IN THE DEVELOPMENT OF PREDICTIONS OF EARTHQUAKES AND VOLCANOES

EARTHQUAKE PREDICTION FROM THE VIEWPOINT OF EARTHQUAKE ENGINEERING

ANTHONY S. KARAMANOS[1]
122 Sirinon Str.
175 62 Paleo Faliro
Athens
Greece

General

During the last decade there has been a considerable development and an intense activity of Earthquake Engineering all over the world. The general objective of that relatively new field of engineering is the study of the behaviour of buildings and other engineering structures (bridges: towers: reservoirs: etc.) under earthquake action. For the attainment of the above general objective, earthquake engineering follows two main directions: one towards the design of new earthquake resistant structures, and the other towards the protection and strengthening of existing structures. In following these two broad directions, earthquake engineering makes use of all available scientific knowledge that can be of help, not only from the related fields of structural and geotechnical engineering, but also from other fields like geology, geophysics and seismology. In this context, an interesting question arises whether prediction of earthquakes, a subject widely discussed at present, can be of use to earthquake engineering in its effort to achieve the above objectives.

Before any attempt to answer the question, it is necessary that the term *prediction,* is considered, its meaning be analysed and its components be described. Prediction of an earthquake means to know its place of occurrence, its time of occurrence, and its magnitude before the earthquake actually happens. According to many authors, prediction can be *long-term* or *short-term* depending on the accuracy of estimation of its time of occurrence, expressed in years for the former and in months or weeks for the latter. Another classification is made to *special* prediction in contrast to the *general* prediction, which is the knowledge of the seismicity of an area.

In older times, the problem of prediction of earthquakes was a subject of astrologers,

[1] **Head of the Earthquake Engineering Department of the Earthquake Planning and Protection Organisation, Athens, Greece.**

J. Nemec et al. (eds.), Prediction and Perception of Natural Hazards, 135–142.

amateur seismologists etc. It is only during the last twenty years that scientific methods were developed, the application of which has shown that the prediction in the aforementioned sense can be a realistic target for the future (Papazachos B. & C., 1989). Therefore earthquake prediction is practised today on a serious scientific basis, and from that point of view it surely has a significant theoretical interest, since it is related to important physical processes in the interior of earth (Papazachos B. & C., 1989). But the importance of prediction is more than that. Prediction has an enormous practical value, since it is associated with the protection of people's lives and property from the catastrophic consequences of earthquakes.

In the following the contribution earthquake prediction either to the design of new earthquake resistant structures or to the protection and strengthening of existing structures is analysed.

Design of new earthquake resistant structures

This direction is followed today by almost all earthquake prone countries by establishing a so called *long-term earthquake planning*. The objective pursued here, is to protect the lives and property of people through design and construction of earthquake resistant buildings and other engineering structures. It is evident that such task belongs to engineers, who must determine with the maximum possible accuracy the seismic actions with which their structures should be designed. In accordance with present knowledge, these actions are determined on a probabilistic basis, considering the fact that an earthquake is an accidental situation. The question is how prediction, as it has been defined previously, can contribute to the determination of the above seismic actions.

In the case under examination, the factor *time* as one of the three components of prediction (and the most important one), has a span of many years, usually equal to the conventional lifetime of ordinary civil engineering structures (e.g. 50 to 60 years). For that span of time, the engineer must know with a certain probability the maximum expected value of the seismic action in the area where his structure is to be built. This value of the seismic action can be expressed in anyone of its parameters such as the ground acceleration, the ground velocity or the macroseismic intensity. For example, it can be said that the maximum expected ground acceleration in some area is 0.21g with a 10% probability of being exceeded in the next 50 years (which means an average return period 475 years),. The same can be said for the other parameters.

The above type of information is what is needed by engineers in order to determine the seismic actions for the design of structures, and is provided by technical seismology in the context of a special kind of long term prediction known as *seismic hazard* estimation. Thus in the case of long-term planning, the main contribution is due to the above special kind of prediction which essentially is reduced to the problem of estimating the seismic hazard. Many countries have solved the above problem on a national basis, and they have hazard maps providing the type of aforementioned information. Greece for example, has seismic hazard maps which divide the whole country into four zones of equal seismic hazard I,II,III,IV (Fig.1). For each zone the maximum expected ground acceleration and velocity are given, in terms of different probabilities and different average return periods as in Fig.2 (Geophysical Departments of Athens and Thessaloniki Universities, 1989).

The information furnished by hazard maps is implemented by the *Seismic Design Codes*, which are the main tool for the design of earthquake resistant structures under an established long-term planning. These codes generally cover the whole area of a country, and they reflect the existing level of knowledge in designing new earthquake resistant structures (Draft of New Greek Seismic Design Code, 1989).

Protection and Strengthening of Existing Structures

This direction is followed today by establishing a so called *medium-term earthquake planning*. The objective pursued here, is the protection of existing structures. This leads to the general problem of vulnerability of structures, one of the most serious problems that big cities and urban centers face today in earthquake prone areas. In general, these cities have extensive zones (historic centers, old districts etc.) with buildings vulnerable to strong earthquakes. This may be due to various reasons, such as bad soil conditions, poor quality of construction, deterioration of materials etc. To the above reasons one must add the disordered expansion of cities due to rapid population increase. Consequently, these cities are extremely vulnerable to a strong earthquake, and in case of such an event very serious problems can be created especially if buildings like hospitals, schools, fire stations, telephone centres, etc., are heavily damaged or collapsed.

Theoretically the above problem could be faced by putting in force a gigantic programme of strength evaluation and retrofit of all vulnerable structures in big areas, but this is practically impossible for two reasons: first for the tremendous amounts of money needed (financing reason), and second, for the lack of knowledge and expertise to handle many different types of structures, constructed in many different periods of time (technical reason).

The problem can be made easier by medium-term prediction which, for some years ahead e.g. one or two decades, can identify smaller areas of high seismic risk enabling thus the decision makers to establish *priorities* in checking and retrofitting existing structures. This kind of prediction, usually called by seismologists long-term prediction, in contrast to the estimation of seismicity, seemed almost impossible about thirty years ago, but in the mean time much progress has been made in developing methods to make that kind of prediction possible (Papazachos B & C, 1989).

Many studies have been made and maps have been published (USA, Japan, Greece etc.), giving predictions of expected strong earthquakes in a certain area and for the next one or two decades. As an example, figures 3 and 4 show two maps of probabilistic prediction in Greece and the surrounding larger area for the period 1986-2006 (Geophysical

Laboratory, University of Thessaloniki, 1986). The first map refers to strong surface earthquakes while the second one to medium-depth earthquakes (depth over 100 km).

It is evident that prediction maps of that kind are very useful for decision makers to fix priorities and prepare programmes for the strengthening of existing vulnerable structures. Present seismic codes provide guidance only for the design of new structures, but they still do not include information for the strengthening of existing structures. Nevertheless, it is certain that in the near future this type of information and guidance will be included in the codes, alongside with the development of the type of prediction in question.

Although it is out of the scope of the present paper, a note must be made on the importance of this type of medium-term prediction for the elaboration of realistic emergency plans in areas where strong earthquakes are expected. A recent characteristic example is the prediction of the LOMA PRIETA earthquake, which occurred on October 17, 1989 in California, which permitted local authorities to take increased emergency measures, and thus to mitigate the effects that the earthquake provoked.

Conclusion

From what has been said above, it is certain that long-term prediction can be of great value and usefulness to earthquake engineering, because it contributes to the design of new earthquake resistant structures through estimation of seismic hazard, and on the other hand because it contributes to the protection and strengthening of existing vulnerable structures, through identification of high risk areas for some years ahead. For this reason, research on that kind of prediction must be intensified so that new results can be used for the improvement of the seismic resistance of structures.

References

Draft of New Greek Seismic Design Code (1989). EPPO.

Geophysical Laboratory, University of Thessaloniki (1986). Long-term Prediction of Earthquakes with Seismic Methods, Thessaloniki.

Geophysical Departments of Athens and Thessaloniki Universities (1989). Geodynamic Institute of Athens Osservatory, Institute of Technical Seismology and Earthquake Resistant Structures. Elaboration of Seismic Hazard Map of Greece, EPPO, Final Report.

Papazachos B. & C., (1989). "The Earthquakes of Greece", Editions Ziti Thessaloniki.

Workshop on Seismic Risk Reduction in the Mediterranean Area - SEISMED (1990). Genoa, Italy

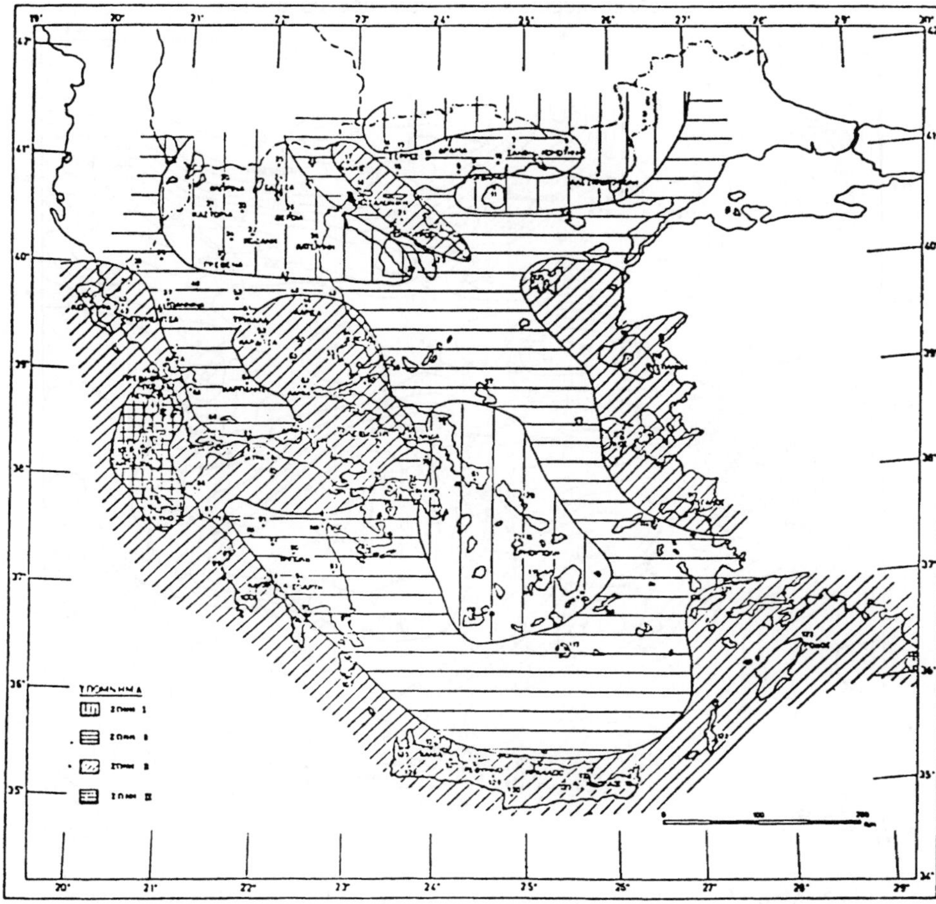

Fig.1 Map of Greece showing the four zones I, II, III, IV of equal seismic hazard. Positions of greater ekistic areas and prefecture capitals are indicated.

By courtesy of Institutions involved and EPPO

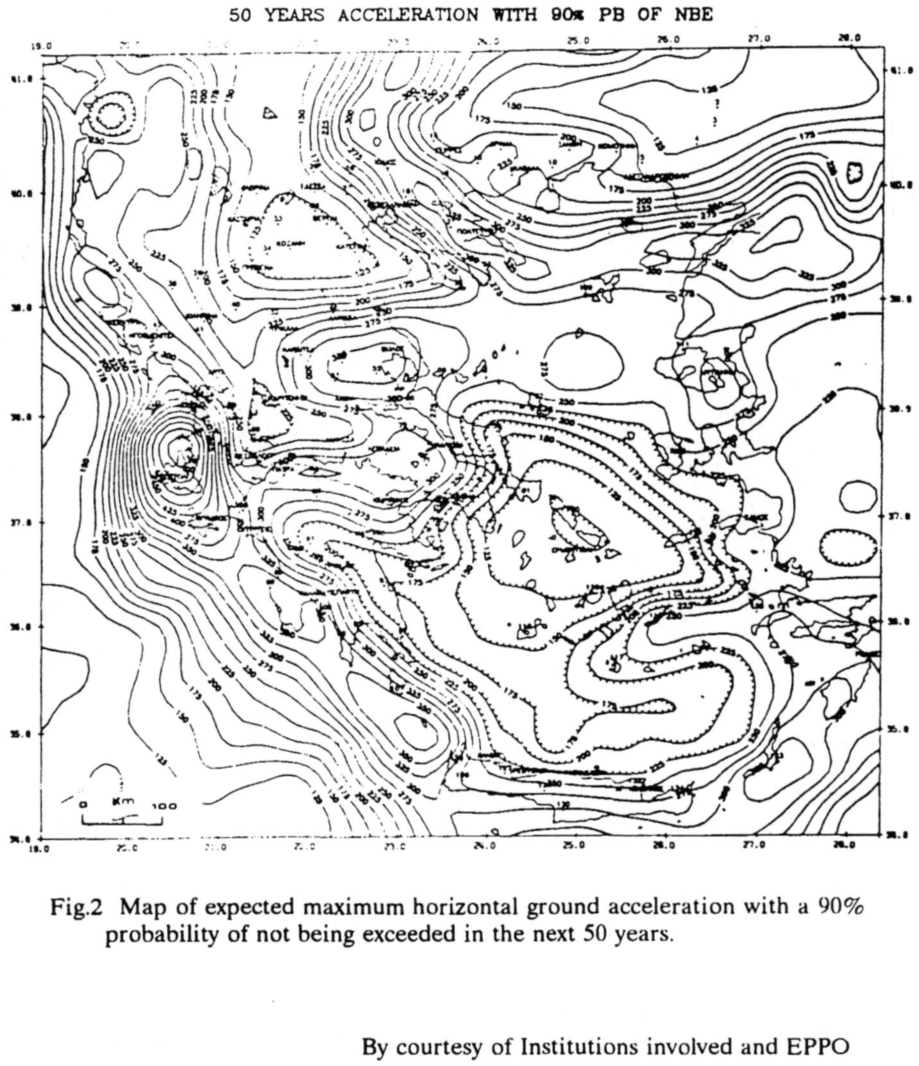

Fig.2 Map of expected maximum horizontal ground acceleration with a 90% probability of not being exceeded in the next 50 years.

By courtesy of Institutions involved and EPPO

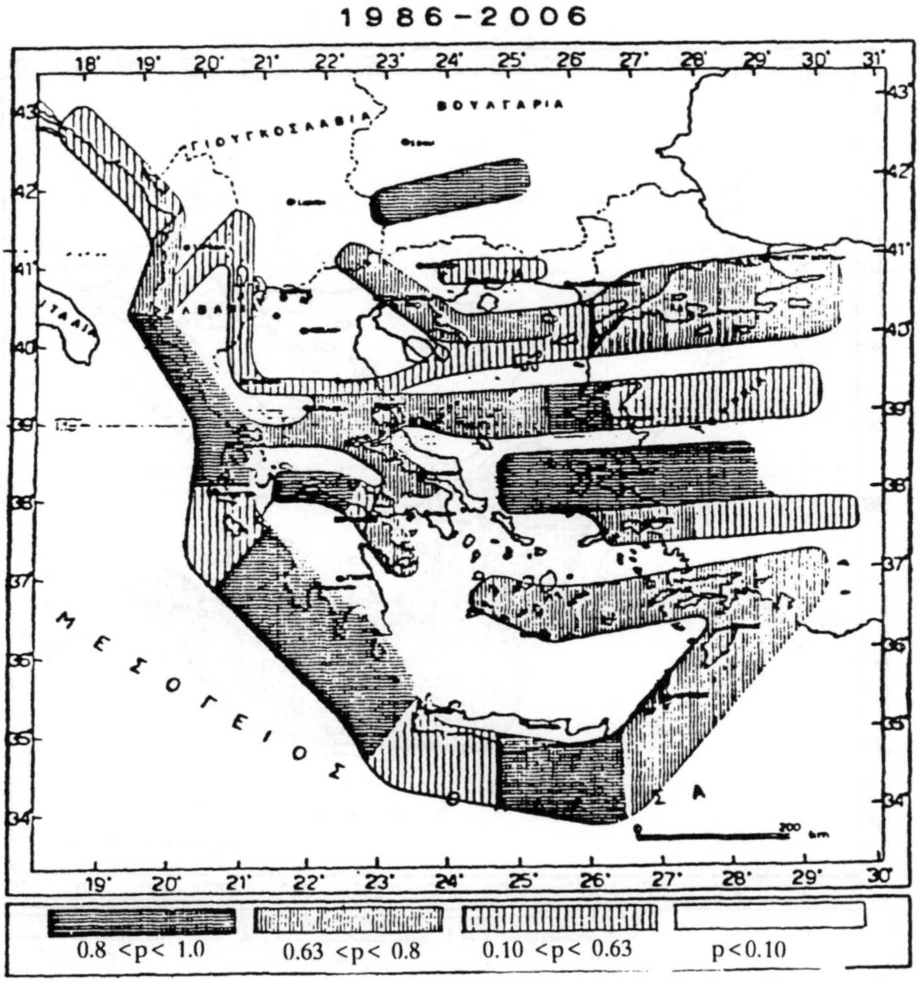

Fig.3 Map of probabilistic prediction of strong shallow earthquakes in Greece and surrounding areas, for the period 1986-2006.

By courtesy of Geophysical Laboratory-University of Thessaloniki and EPPO

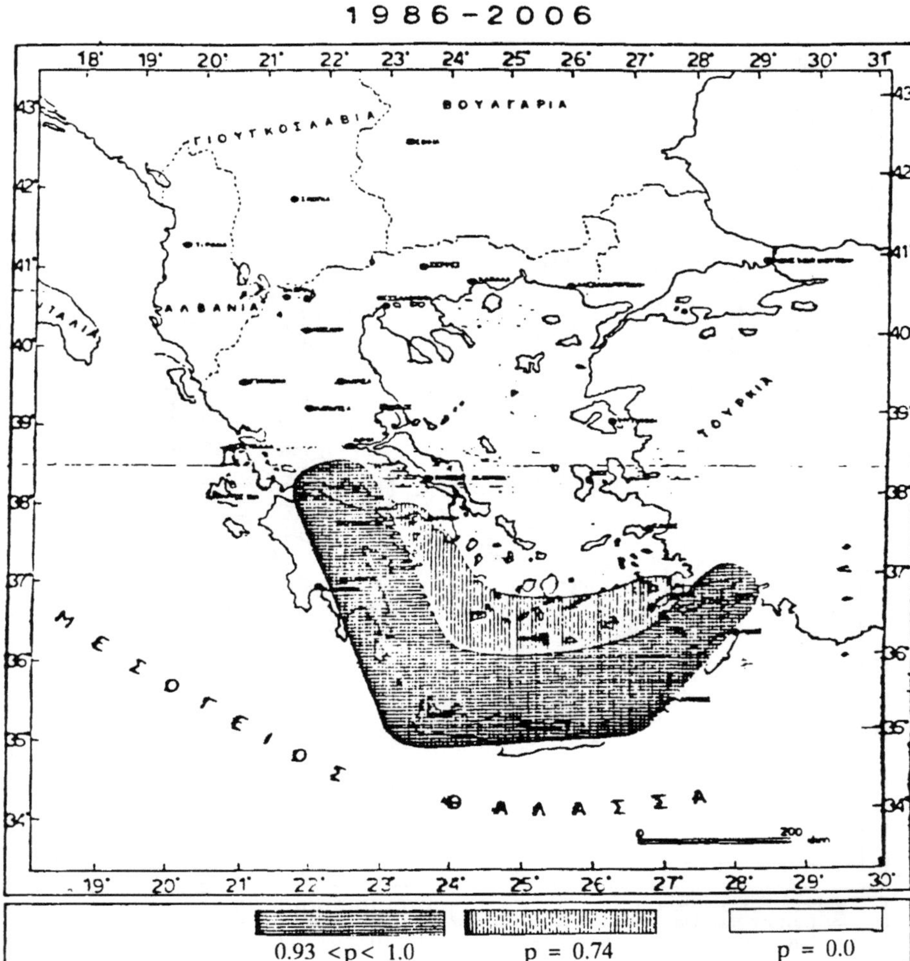

Fig.4 Map of probabilistic prediction of strong medium-depth earthquakes in Greece and surrounding areas, for the period 1986-2006.

By courtesy of Geophysical Laboratory-University of Thessaloniki and EPPO

COMMUNICATING PUBLIC EARTHQUAKE RISK INFORMATION

DENNIS S. MILETI
Colorado State University
Department of Sociology
C242 Clark Building
Fort Collins, Colorado 80523
U.S.A.

Efforts in the United States continue to communicate information about earthquake risk to the public. These efforts fall into two categories. First, risk is communicated to inform and educate citizens about general earthquake risk. Second, risk is communicated in reference to the prediction of warning of a specific impending earthquake. Central to both risk communication types are the perceptions which the public forms about earthquake risk since these perceptions direct public behavior.

A THEORETICAL MODEL

Existing research on public risk communication suggests that risk perception is a process that is explained by aspects of how risk is communicated, and by variation in the characteristics of the people who receive those communications. These conclusions are reviewed and then briefly summarized in a theoretical model in this section of this paper.

Public perception of risk as a social process

Findings from research on communications about earthquake risk suggest that public perception of risk is a multi-dimensional concept, and that it is more appropriately seen as a social process than as a single variable. This process contains factors such as the understanding or the meaning people impute to heard risk communications, their belief in what was communicated, and whether or not the communicated risk is personalized versus seen as someone else's problem.

Communicated risk information must be understood. This does not mean understanding what is heard, but rather the personal attachment of meaning to the message. Meaning or understanding can vary among people, and these varied understandings may or may not conform to the meaning intended by those who issue the message. For example, a fifty

J. Nemec et al. (eds.), Prediction and Perception of Natural Hazards, 143–152.

percent earthquake probability may be interpreted as certainty by some or unlikely by others. A singular risk message could be understood as involving no risk by some but as a threat to human life by others. People attach different meaning to a risk message. Once the risk information is understood, it is necessary for people to believe the contents of the message. Belief in the message can vary among a public at risk. Once people believe the warning, they consider its implications for themselves. If people do not think that risk information was meant for them, they likely will ignore it. If people think they are the intended targets of emergency information, they may act accordingly. When a person has heard the risk information, formed an understanding of whatis being said, believed what is being said, and established the belief that he or she will be personally affected, that person must then decide what to do. Response to risk communications is not the result of a single decision after hearing a risk communication. It is instead the eventual consequence of a social psychological process out of which emerge perception of risk and ideas about what to do because of it.

Members of the public typically go through each of these steps outlined above (understand-believe-personalize) each time that a new risk communication is received. The resulting perception of risk which they form consequently evolves over time. It results from the process of interacting with risk information over multiple communications. Additionally, people do not passively await the arrival of more information. Instead, most people actively seek out additional information. When risk information is received, most people try to verify what they heard by seeking out more information from another source or person. This seeking of new information to confirm prior information, or receiving new information which confirms prior information, has typically been referred to as risk information confirmation.

Factors that influence public perception of risk

The principle that people behave in response to risk communications in ways consistent with their perceptions of risk has been researched in a variety of natural and technological emergencies and does not rest only on the findings of earthquake studies. This research has been of various types; some studies have been descriptive, while others tested hypotheses. Nevertheless, enough evidence exists to conclude which are the characteristics of risk communications that maximize the profitability that they will be correctly understood, believed, personalized and acted upon by the public. The most effective communications are those who are, for example specific about impact location, give protective action recommendations for the public to take, the time to impact, and the character of risk. In addition, the most effective communications are consistent and certain, address why they should be acted on, delivered through multiple channels of communication, repeated frequently rather than just said once and labeled as coming from a panel of officials, scientists, and experts credible for everyone. These sorts of factors illustrate variation in the substance of the risk communication itself. Clearly, risk communication that is effective is itself a process and not a singular act.

The public, however, is comprised of very different kinds of people. Variations among people also have been documented to affect the risk perceived and actions taken in response to public communications about earthquake risk. For example, Research evidence suggests that risk is easier to perceive if the person receiving a communication observes cues in the environment that supports what is being said, is close to the projected area of impact, has higher socioeconomic status, has experience with a damaging earthquake in their personal history and others.

A theoretical summary

In sum, risk communication factors (e.g. source, consistency, channel of communication, accuracy, clarity, certainty, guidance about what to do, specificity about risk location, frequency of repetition and so on) and factors that differentiate between members of the public (for example, experience, socioeconomic status, environmental cues and others) interact to determine the risk which people perceive (understanding, belief, and risk personalization) which then directs public response to risk communications.

This "theory" was recently subjected to a comprehensive test by examining public response to the first scientific earthquake prediction in the United States that has ever been approved by both the National and California Earthquake Prediction Evaluation Councils. The following sections of this paper describe that prediction, how the theoretical model of earthquake risk communication was tested, and the study results.

PUBLIC RESPONSE TO RISK COMMUNICATIONS ABOUT THE PARKFIELD EARTHQUAKE PREDICTION EXPERIMENT

The prediction

The Parkfield Earthquake Prediction was for a moderate earthquake of magnitude 5 to 6 on the Parkfield segment of the San Andreas fault between 1985 and 1993. It was given a 95 percent probability of occurrence. Contained in the prediction was the possibility that the next Parkfield earthquake could be 7 in magnitude. The larger event, were it to occur, could result in significant damage. The director of the U.S. Geological Survey issued a public statement on April 5, 1985. The release of this prediction was a national medial event. The prediction was reported in national weekly news magazines. The national media came to interview the residents of Parkfield. The interviews and editorial comments were broadcast on national network evening news programs.

California's Governor's Office of Emergency Services (OES) prepared and mailed a brochure describing the prediction and recommended actions to more than 122,000 central California households--every household within the predicted area at risk. The brochure described the Parkfield Earthquake Prediction itself, advised people about how to prepare for the earthquake, described the effects that could be expected from the predicted earthquake and advised people about how to get additional information.

The study on public response

The purpose of the study was to find out what the public thought and did in response to risk communications about the prediction, why they thought and did it, and to comprehensively test the theory of risk communication induced from other research. We picked three communities to study within the area-at-risk from the predicted earthquake. We wanted to be able to compare earthquake prediction response between a community that had recent experience with those without it, and response in a community close to the predicted epicenter with that in a community far away. Data were collected in each of these communities. Questionnaires were mailed to random samples of households with subsequent follow-ups by mail to increase the number of responses. A total of 948 completed questionnaires were received from the three communities.

Study findings

We found no difference in public perception or response to the earthquake prediction by demographic characteristics usually included in studies, such as age, income, educational levels, and so on. Instead, we found differences because of the following personal characteristics. Response was greater among those: 1) with recent damaging earthquake experience, 2) whose friends and neighbors took preparedness and mitigation action; and 3) among people for whom the hazard was already salient enough for them to have already taken protective action against earthquake risk prior to the Parkfield earthquake prediction.

We found that perceived characteristics of the set of information received (not simply the brochure itself, but all remembered information about the earthquake prediction and what to do about earthquakes) affected people's response to the prediction. People were more likely to feel that they had "heard of" the Parkfield earthquake prediction; to understand, believe, and personalize the risk from it; and to have taken protective action the more they:

- perceived the full array of received prediction messages as basically consistent with each other

- remembered the predicted earthquake's magnitude, potential for damage, and other characteristics, which made the prediction seem more certain to them

- remembered specific guidance for appropriate protective actions to take, such as strapping down the water heater and setting up an emergency supplies stockpile

- remembered that they had received these messages through numerous vehicles or channels, and especially that they had received them through the printed word in the mailed brochure and newspapers

- remembered that they received these messages from numerous information sources including scientists and relatives, and especially that they had received them from official sources.

Through a statistical technique called "path analysis," we were able to shed more light on the response process. The fact that this process was found to be identical in all three study communities lends evidence that we can describe a consistent response pattern with confidence. This response pattern represents the most significant influences that occur sequentially in the communication and response process--the factors that count the most.

The first of these is reinforcement. We have seen how risk communication is most effective when it is in a written form. However, to be most effective, the written prediction has to be reinforced by multiple messages delivered through multiple channels and from, importantly, official sources, as well as from other sources. A second way that the prediction is reinforced is by individuals observing others taking preparedness and mitigation action.

The direct next stop toward action is taken when the individual begins to seek additional information about the prediction, the risk, what to do about it, and what others are doing about it. This information search occurs when members of a community interact to discuss the earthquake hazard and how to deal with it. The process results in the person forming his or her individual definition of the risk and pictures about how to proceed--what actions to take to mitigate and prepare. This process of coming to "own" the risk information is essential to taking protective action.

In summary, then, the key sequential reasons underlying the taking of protective actions are:

- communicated public risk information that is printed and reinforced.
- searching for more information through interacting with others
- coming to "own" the earthquake prediction information and the appropriate actions to take in response to it.

IMPLICATIONS FOR COMMUNICATING EARTHQUAKE RISK INFORMATION TO A PUBLIC

The findings from this study have implications for those responsible for informing the public that an earthquake is likely to occur, and that they are at risk. Doing this is more simple than we had originally thought and recommendations follow.

1. Go to the public with a written brochure and, if funding exists, mail it to their homes. If insufficient funding exists for mailing, publish it in area newspapers. The brochure will work best if it states that it is from official government sources. The brochures should explain specially: a) what the risk is, b) where it is going to happen, c) when it is going to happen, d) what the effects will be, e) what people should do before, during and after it, and f) where to get information about it. This information should be as clear and certain as possible.

2. Mailing a brochure is not enough--it must be supplemented. The public needs to be targeted with the message from as many different sources and through as many different channels as possible. This supplemental "barraging" makes the mailed brochure effective. Therefore, go to the media with the same and consistent supplemental information to that appearing in the brochure.

3. People need multiple information sources to reinforce the risk information in the brochure. Reinforcement also occurs in other ways, such as by people seeing neighbours, friends, and relatives preparing for earthquake risk. Consider visible demonstration projects in communities that are targets for earthquake prediction to reinforce the information.

4. Capture peoples attention, spark their interest, and have them begin considering that they should do something about the risk. Despite the fact that they receive the brochure, they will not immediately take action. Instead, they will discuss the risk at local organisations and talk with their friends and neighbors about it. This process permits them to gather information and induce their own ideas about the level of risk and what they should do about it. They may need to feel that taking some protective action is their own idea. This process of information "ownership" takes time. Taking preparedness and mitigation action follows as the result of this process, rather than as the result of the mailed brochure.

5. This being the case, position information in the local community for use during this "milling around" process, such as coloring books, brochures, slide shows, film strips, classes on earthquake preparedness and mitigation, consultation on family and work place earthquake plans, and so on.

6. It might be better not to name earthquake predictions after towns (like Parkfield) because this may perceptually limit the perception of risk in towns distant from the epicenter of the predicted earthquake.

REFERENCES

Journal articles

Anderson, W. (1969) "Disaster Warning and Communication Processes in Two Communities." The Journal of Communication 19 (June):92-104.

Baker, E. J. (1979) "Predicting Response to Hurricane Warnings: A Reanalysis of Data from Four Studies." Mass Emergencies 4(1):9-24.

Dillman, D. A., Schwalbe, M. L. and Short, J. F. (1983) "Communication Behavior and Social Impacts Following the May 18, 1980 Eruption of Mt. St. Helens." In Mt. St. Helens Three Years Later, Proceedings of a Conference, Pullman Washington, Washington State University.

Drabek, T. E. (1969) "Social Processes in Disaster: Family Evacuation." Social Problems 16(Winter):336-349.

Drabek, T. E. (1983) "Shall We Leave? A Study of Family Reactions When Disaster Strikes." Emergency Management Review 1(Fall):25-29.

Drabek, T. E. and Boggs, K. (1968) "Families in Disaster: Reactions and Relatives." Journal of Marriage and the Family 30 (August):443-451.

Drabek, T. E. and Stephenson, J. S., III. (1971) "When Disaster Strikes." Journal of Applied Social Psychology 1(2):187-203.

Greene, M. R., Perry, R. W. and Lindell, M. K. (1981) "The March 1980 Eruptions of Mt. St. Helens: Citizen Perceptions of Volcano Threat." Disasters 5(1):49-66.

Lachman, R., Tatsuoka, M. and Bonk, W. (1961) "Human Behavior During the Tsunami of May, 1960." Science 133(May 5):1405-1409.

Lindell, M. K. and Perry, R. W. (1987) "Warning Mechanisms in Emergency Response." International Journal of Mass Emergencies and Disasters 5(2):137-153.

Mileti, D. S. and Sorensen, T. H. (1988) "Planning and Implementing Warning Systems." Mental Health Care In Mass Emergencies: Theory and Practice, edited by M. Lystad, New York:Brunner/Mazel Psychological Stress Series, 321-345.

Perry, R. W. (1979a) "Evacuation Decision-Making in Natural Disasters." Mass Emergencies 4(March):25-38.

Perry, R. W. (1979b) "Incentives For Evacuation in Natural Disaster-Research Based Community Emergency Planning." Journal of the American Planning Association 45(4):440-447.

Perry, R. W. and Green, M. R. (1982) "The Role of Ethnicity in the Emergency Decision-Making Process." Sociological Inquiry 52(Fall): 309-334.

Perry, R. W., Lindell, M. K. and Green, M. R. (1982) "Threat Perception and Public Response to Volcano Hazard." Journal of Social Psychology, 116(April):199-204.

Quarantelli, E. L. (1984) "Perceptions and Reactions to Emergency Warnings of Sudden Hazards." Ekistics 309(Nov-Dec):511-515.

Scanlon, J. and Frizzell, A. (1979) "Old Theories Don't Apply: Implications of Communication in Crisis." Disasters, 3:315-319.

Sorensen, J. H. and Gersmehl, P. J. (1980) "Volcanic Hazard Warning System: Persistence and Transferability," Environmental Management 4(March):125-136.

Sorensen, J. H. and Mileti, D. S. (1989) "Warning and Evacuation: Answering Some Basic Questions." Industrial Crisis Quarterly 2:136-171.

Turner, R. H. (1983) "Waiting for Disaster: Changing Reactions to Earthquake Forecasts in Southern California." International Journal of Mass Emergencies and Disasters 1(2):307-334.

Withey, S. B. (1962) "Reaction to Uncertain Threat." Man and Society in Disaster, edited by G. Baker and D. Chapman, New York: Basis Books, 92-123.

Book references

Drabek, T. E. (1986) Human System Response to Disaster: An Inventory of Sociological Findings. New York, NY: Springer Verlag.

Friedsam, H. J. (1962) "Older Persons in Disasters." Man and Society in Disaster, edited by George W. Baker and Dwight W. Chapman, New York, NY: Basic Books, 151-184.

Grunfest, E. C. (1978) What People Did During the Big Thompson Flood. Department of Geography, University of Colorado, Boulder, CO.

Mileti, D. S., Drabek, T. E. and Haas, J. E. (1975) Human Systems in Extreme Environments: A Sociological Perspective. Institute of Behavioral Science, University of Colorado, Boulder, CO.

Perry, R. W. and Green, M. R. (1983) Citizen Response to Volcanic Eruptions: The Case of Mt. St. Helens. New York, NY: Irvington Publishers, Inc.

Perry, R. W. and Lindell, M. K. (1986) Twentieth-Century Volcanicity at Mt. St. Helens: The Routinization of Life Near an Active Volcano. School of Public Affairs, Arizona State University, Tempe, AZ.

Perry, R. W. and Mushkatel, A. (1984) Disaster Management: Warning Response and Community Relocation. Westport, CT: Quorum Books.

Perry, R. W. and Mushkatel, A. (1986) Minority Citizens in Disasters. GA: University of Georgia Press.

Perry, R. W., Lindell, M. K. and Green, M. R. (1981) Evacuation Planning in Emergency Management. Lexington, MA: Lexington, Books.

Quarantelli, E. L. (1980) Evacuation Behavior and Problems: Findings and Implications From the Research Literature. Disaster Research Center, The Ohio State University, Columbus, OH.

Rogers, G. O. (1985) Human Components of Emergency Warning. University Center for Social and Urban Research, University of Pittsburgh, Pittsburgh, PA.

Sorensen, J. H. (1987) "Warning Systems in the 1985 Cheyenne Flash Flood." What We Have Learned Since the Big Thompson Flood. Gruntfest, E. (ed.). Institute of Behavioral Science, University of Coloradlo, Boulder, CO.

Turner, R. H., Nigg, J. M., Heller, D. and Young, B. S. (1984) Wwiting for Disaster. University of California Press, Berkeley, CA.

Turner, R. H., Nigg, J. M., Heller, D. and Young, B. S. (1981) Community Response to Earthquake Threats in Southern California. University of California Press, Los Angeles, CA.

Turner, R. H., Nigg, J. M., Heller, D. and Young, B. S. (1979) Earthquake Threat. Institute for Social Science Research, University of California, Los Angeles, CA.

Williams, H. B. (1964) "Human Factors in the Warning and Response Systems." The Threat of Impending Disaster, Grosser et al., (eds.), Cambridge, MA: MIT Press, 79-1047.

Withey, S. B. (1962) "Reaction to Uncertain Threat." Man and Society in Disaster, Baker, G. and Chapmen, D. (eds.) Chapman, NY: Basic Books, 92-123.

Technical reports

Leik, R. K., Carter, T. M. and Clark, J. P. (1981) Community Response to Natural Hazard Warnings: Final Report. Minneapolis, MN: University of Minnesota.

Rogers, G. O. and Nehnevajsa, J. (1984) Behavior and Attitudes Under Crisis Conditions. Washington, DC: FEMA.

Sorensen, J. H., Vogt, B. M. and Mileti, D. (1987) Evacuation: An Assessment of Planning and Research. Oak Ridge National Laboratory Report ORNL-6376.

UNCERTAINTIES IN THE ESTIMATE OF STRONG GROUND MOTION IN THE SURROUNDINGS OF A LARGE EARTHQUAKE

GIULIANO F. PANZA[1] & PETER SUHADOLC[1]
Institute for Geodesy and Geophysics
University of Triest
Via Dell'Università 7
34100 Trieste
Italy

PAOLO HARABAGLIA
Massachussets Institute of Technology
Cambridge, MA
USA

INTRODUCTION

The engineering problem of understanding the collapse of a building during a major earthquake is greatly eased if the actual seismic motion on the site is known. Whenever a sufficient number of strong motion records is available, as in the case of the november 23, 1980 Irpina (Italy) earthquake, we can realistically simulate it for frequencies up to 1 Hz using "complete" synthetic seismograms. In our case we apply a normal mode summation technique. Through a simple minimization process we obtain a source model that reproduces the observed records. Then we use it to estimate the seismic ground motion at other sites. We also investigate the stability and uncertainties in our solution.

Modeling of observed signals

The fault is modeled as a grid of point sources: seismograms are given by the superposition of the signals due to each point source activated during the rupturing process. Moreover, the fault is approximated with a plane or a set of planes and we assume that the direction of slip along each plane does not vary appreciably. Therefore our problem is to find the spatial orientation of the fault plane(s) and its rupturing history. This is tantamount to determine the evolution of the rupture front both in space and time and to find the seismic moment released on each portion of the fault. In our case this means finding a weight and an origin time for each point source. To simplify further the problem, we assume a 1-D layered structural model around the source volume and a locally

[1] Also at: International Centre for Earth and Environmental Sciences, Miramar, Italy

J. Nemec et al. (eds.), Prediction and Perception of Natural Hazards, 153–158.

constant rupturing velocity. These approximations might seem quite strong, but we have deliberately chosen to use a relatively simple theoretical source model that is still able to produce realistic ground motion estimates. If S_{ij} is the contribution of the i-th point source to the record A_j, we look for a solution that minimizes the expression

$$R = \frac{\sum_j \left[A_j - \sum_i w_i S_{ij}\right]^2}{\left[\sum_i w_i^2\right]^{1/2}}, \ (i=1,\dots,M; j=1,\dots,N) \tag{1}$$

where w_i is the "weight" related to the seimic moment, given to the i-th point source. M is the total number of point sources which give rise to synthetic signals, N is the total number of records used in the minimization of R and $[\Sigma_i w_i^2]^{1/2}$ is also a normalization term such that the number of significant point sources is kept as low as possible. The minimization is performed through an iterative procedure that makes use of the conjugate gradient method based on analytically computed derivatives of R with respect to w_i.

The key to a successful minimization lies in the ability to reproduce all the phases of a signal and not only the first arrivals. For this reason we base our strategy on the mode summation method that gives "complete" seismograms, provided a sufficient number of higher modes is summed.

The 1980 Irpinia, Italy, earthquake

The method is tested for the November 23, 1980 Irpinia, Italy (Ms=6.9) event. This earthquake has been studied by many researchers. We have selected six of the strong motion accelerographs of the ENEA-ENEL network operating in the earthquake area so as to have a single structural model. The small number of stations is still sufficient to invert for slip on preassigned geometry. We also limited ourselves to the vertical components, low-pass filtered with a Gaussian filter having a cutoff frequency of 1Hz, since they are usually less affected by local conditions. Moreover, since the ENEA-ENEL accelerographs did not record an absolute time, appropriate triggering times are inferred.

The final source geometry is shown in Fig. 1. The rupturing mechanism giving the best match between observed and synthetic seismograms is shown in Fig. 2a, whilst the observed and synthetic seismograms are shown in Fig. 2b. We want also to remark that since we are interested in an engineeristic application, we have limited ourselves to look only for solutions similar to those already available in literature. Therefore this model is the result of the evaluation of the slip distribution on some hundred different fault geometries centred about those models, and actually consists of about 1500 point sources spaced by 1 Km.

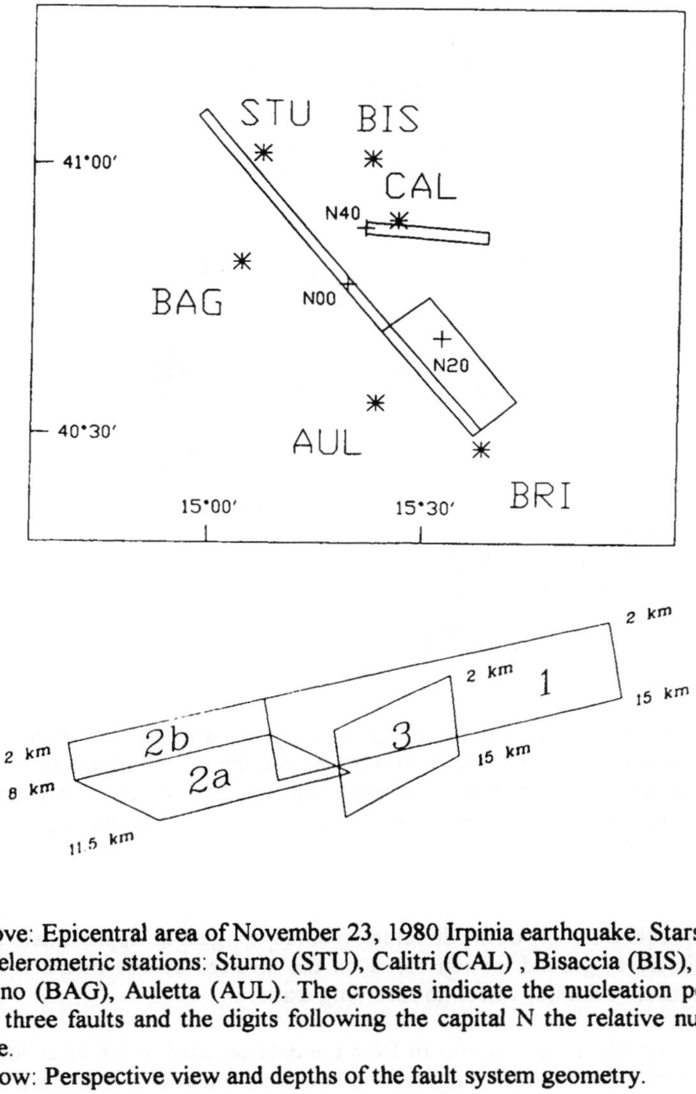

Fig. 1: Above: Epicentral area of November 23, 1980 Irpinia earthquake. Stars denote accelerometric stations: Sturno (STU), Calitri (CAL) , Bisaccia (BIS), Bagnoli Irpino (BAG), Auletta (AUL). The crosses indicate the nucleation points on the three faults and the digits following the capital N the relative nucleation time.
Below: Perspective view and depths of the fault system geometry.

Estimate of ground motion where observations are not available

Once we got our "best" source model, we alternatively discard one of our records and perform again the minimization using the remaining ones. By comparing them with the "best fit" model, we have a measure of the stability of our solution. Moreover, if we solve

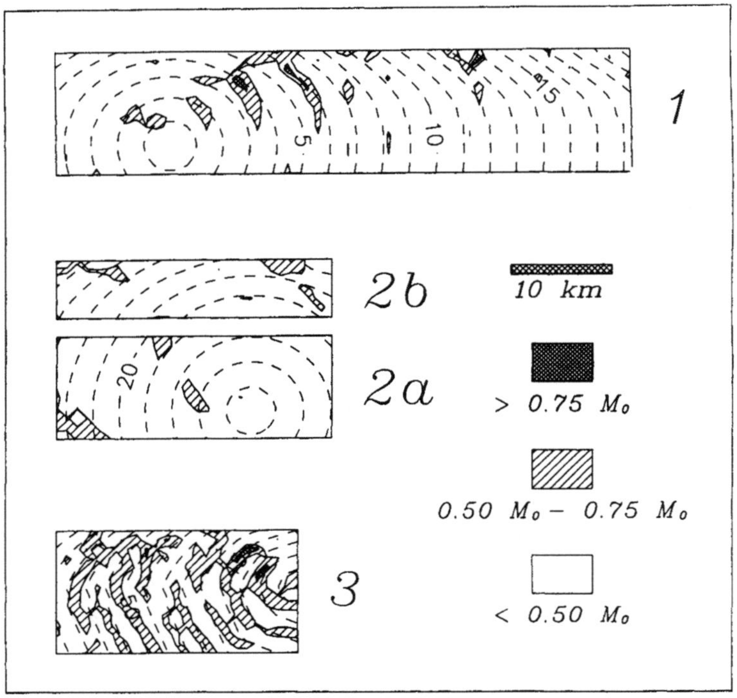

Fig. 2: (a) Source Model (distribution of moment release) along the faults. The moment is normalised to 100 by the maximum estimated value (3×10^{23} dyne-cm). Dashed lines represent the rupture front at 1s intervals. The total seismic moment released is 1.13×10^{26} dyne-cm, and the total average slip is 35 cm.

the forward problem for any of the discarded records, we gain an approximate idea of how good an estimate of ground motion based on the inverted source model is at localities around the causative fault(s) where no recording exists.

Alternatively, we take away one record from the data set used in the inversion procedure and we perform the minimization again. The new source model is then used to solve the forward problem for the excluded record. In Fig. 3a we show, as an example, the source model obtained without the contribution of Bagnoli and both observed and the theoretical

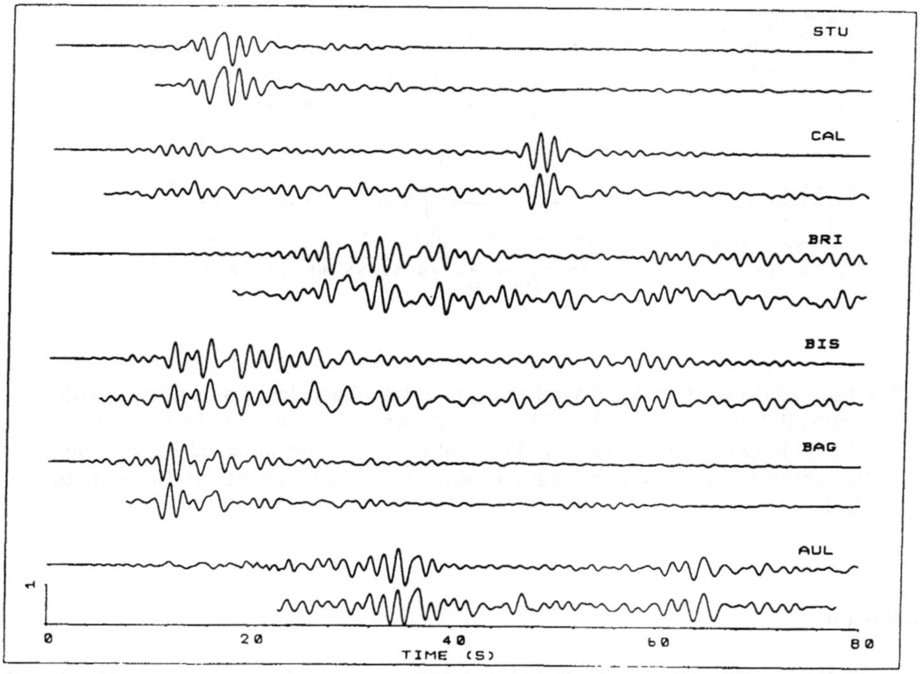

Fig. 2b: Synthetic accelerograms (above) and observed ones (below) filtered with a cut-off frequency of 1 Hz. They are shown normalised to their maximum amplitude (expressed in cm/s^2) - AUL: 8.4 (Synthetic), 7.8 (Observed); BAG: 32.7 (S), 36.3 (O); BIS: 26.5 (S), 32.2 (O); BRI: 7.6 (S), 9.3 (O); CAL: 59.2 (S), 70.4 (O); STU: 60.1 (S), 48.7 (O);

signal. In Fig. 3b the same is accomplished for Calitri. Both models are fairly similar to that of Fig. 2a. A similar result holds for Bisaccia. Conversely, taking away Auletta, Brienza or Sturno does not lead to any appreciable result. Therefore, for a successful estimate, the network of recording stations has to be in a position to detect all the main subsources at more than one station, otherwise the ability to realistically estimate the strong ground motion at sites between the stations will be severly hampered.

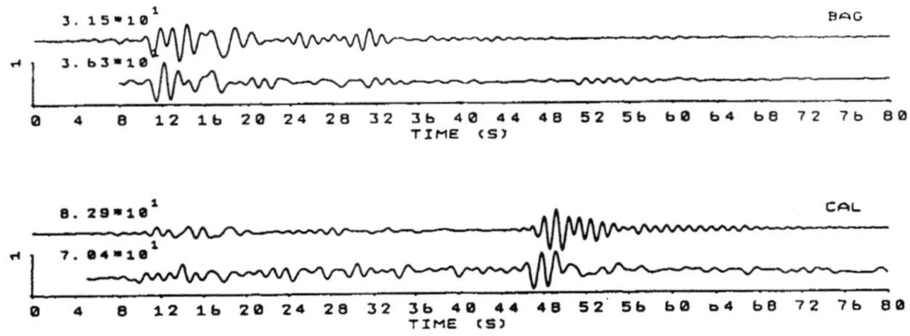

Fig. 3: Observed (lower trace) and synthetic (upper trace) accelerograms filtered with a cut-off frequency of 1 Hz. They are shown normalised to their maximum amplitude (expressed in cm/s^2). The synthetic ones are obtained by inverting the observed records at all stations shown in Fig.1 except: a) Bagnoli; b) Calitri

Conclusion

Due to the introduced approximations, the obtained source model is a gross approximation to the real one. However, since it leads to realistic estimates of the recorded signals and it is basically in accord with the simpler models proposed by other authors it can be considered as sufficiently good and realistic for engineering purposes.

We conclude that a relatively simple source model, a sufficiently dense recording network and the possibility to compute complete synthetic signals permits realistic estimates for frequencies below 1 Hz, of strong ground motion in the area around a big earthquake.

Acknowledgements

This work has been financed by the Ente Nazionale Energie Aternative - Contract No. 3972 (serie III) and by the Consiglio Nazionale delle Ricerche in the frame of the project Gruppo Nazionale per la Difesa dai Terremoti under funds 89.02015.54 and 89.02034.54.

RISK AREA RESIDENTS' CHANGING PERCEPTIONS OF VOLCANO HAZARD AT MT. ST. HELENS

MICHAEL K. LINDELL
Michigan State University
East Lansing, Michigan 48824
U.S.A.

RONALD W. PERRY
Arizona State University
Tempe, Arizona 85287
U.S.A.

Volcanic activity resumed at Mt. St. Helens on March 20, 1980, after 123 years of quiet. Over a period of five days, the threat of a major eruption increased and subsided. Media coverage of the volcano was initially intense, but gradually declined and nearly disappeared by the second week of May (Earle, Southwick and Lindell 1981). In addition to fluctuations in the severity of the threat, there were uncertainties about the areas at greatest risk. During the initial emergency period, the Lewis River Valley on the south side of the mountain was thought likely to be the area of most severe impacts. Later the formation of a bulge on the northwest side of the volcano diverted attention to the Toutle River Valley, which ultimately bore the brunt of the catastrophic explosion on May 18th. Since that time, steam, ash and effusive (dome building) eruptions have decreased, although the threat of flooding caused by ash deposition in the river continued until the very recent completion of a sediment retention dam (Perry and Lindell 1990a).

The volcanic activity at Mt. St. Helens provides an opportunity to study changes in local residents' perceptions of the volcano hazard and their altered patterns of adjustment to its consequences. Understanding public perception of natural hazards has long been recognized as an important element in enhancing the process of hazard adjustment (Burton, Kates and White 1978; Drabek 1986; Mileti 1980). Public risk perception is a problem because risk area residents often have inaccurate beliefs about the hazard agent and its impacts, are unaware of the available adjustments, and may have erroneous beliefs about the advantages and disadvantages of adjustments they do know about (Burton, Kates and White 1978; Slovic, Kunreuther and White, 1974). Previous investigations indicate that adaptive actions are motivated by awareness of the hazard, frequent thought about it, knowledge about how it can impact the community, and feelings of personally vulnerability to the potential consequences of hazard impact (e.g., Janis and Mann 1977;

J. Nemec et al. (eds.), Prediction and Perception of Natural Hazards, 159–166.

Kunreuther, Ginsberg, Miller, Sagi, Slovic, Borkan and Katz 1978; Mileti and Sorensen 1987).

Data on local residents' perceptions and adjustments to the volcano hazard at Mt. St. Helens were collected in a series of studies, two of which are addressed here. The first ("acute threat") study was conducted during the first week of April, 1980, immediately following the Governor's declaration of a state of emergency. Greene, Perry and Lindell (1982) immediately following the Governor's declaration of a state of emergency. Greene, Perry and Lindell (1982) interviewed residents of three communities south and west of the volcano (Cougar, 15 miles; Woodland, 25 miles, and Longview, 40 miles). The second ("chronic threat") study, conducted in 1983, involved an assessment of risk area residents' long term perceptions and adjustments. Toutle (the community most severely affected) was the focal site, with Lexington, a small community downriver from Toutle and just north of Longview, the comparison (Perry and Lindell 1990a).

Acute threat: risk perception

Prior to the eruption of Mt. St. Helens, research addressing perception of volcano hazards had been conducted on those living in areas close to the Hawaiian volcanoes of Kilauea and Mauna Loa. The findings indicate local residents are knowledgeable, but relatively unconcerned about the hazard. Murton and Shimabukuro (1974) and Hodge, Sharp and Marts (1979) found that the inhabitants of these areas did not consider the volcanoes to be hazards to human safety, although they did recognize the existence of a threat to property.

In the Pacific Northwest, only geologists seemed concerned about the Cascade volcano hazard prior to the March 1980 eruptions of Mt. St. Helens (e.g., Crandell, Mullineaux and Rubin 1975). Approximately one-third of the respondents in each of the communities of Cougar, Woodland and Longview admitted that they were completely unaware of the potential for volcanic activity in their area prior to the initial eruptions. However, the March 1980 ash and steam eruptions did produce a significant change in perceptions of personal risk from volcanic activity. Greene, et al. (1981) reported that the overwhelming majority of the respondents at all three sites thought there was no risk prior to March 1980. By April 1980, however, there was a significant increase in the perceived severity of the volcano threat and the change in perceived risk increased with proximity to the volcano.

Though the overall perception of threat increased, this was not necessarily based upon increased knowledge of the hazard. When asked in a free-response question to name the ways in which their safety would be affected by Mt. St. Helens, 31 percent of the respondents could not identify a specific threat. In addition, there were differences in the residents' concerns from one site to another. Sixty-five percent of the Cougar residents identified ashfall as the principal danger, while most of the remainder (29 percent) of this group mentioned mudflows and floods. By contrast, Woodland and Longview residents were more concerned about mudflows and floods, a category which was named by 70 percent of the respondents, than by ashfall, mentioned by 26 percent of the respondents. Two other possible threats, lava flow and explosive eruptions, were mentioned by at most one or two respondents at each site.

The differences in the salience of these threats can be attributed to differences in the perceived vulnerability of the locations. The pattern of threat salience is similar to the pattern of responses to a question in which respondents were asked to rate the severity of the impacts of ash and mudflows/floods. Seventy percent of the Cougar respondents feared moderate to severe damage from ashfall, while this level of impact was expected by only about 42 percent in Woodland and 23 percent of those in Longview. With respect to mudflows and floods, only 14 percent of the Cougar respondents reported believing that these threats would produce moderate to severe damage, compared to 38 percent of those in Longview and 52 percent of those in Woodland.

A measure of the respondents' perceptions of the severity of the threat to personal safety can be determined from their estimates of the likelihood that they might have to evacuate to protect themselves. Overall, few (17 percent) of the residents of any of the three sites thought it likely that they would even have to evacuate because of the volcano. However, Cougar residents were almost evenly split as to whether the likelihood could best be described by "even odds" (39 percent) or "unlikely" (52 percent). The comparable figures for Woodland residents were 30 percent and 54 percent, respectively, while those for Longview were about 17 percent and 69 percent respectively. Thus, residents from Longview were significantly more confident than those from Cougar or Woodland that they would not have to evacuate.

Acute threat: information seeking

The perceived severity and unfamiliarity of the threat to those exposed (cf. Slovic, Fischhoff and Lichtenstein 1980), when coupled with the immediacy of the hazard, would be expected to lead local residents to engage in an active and extensive search for information. The severity and immediacy of the threat would be likely to stimulate frequent search, while the novelty of the threat would cause informal sources of information (friends, relatives, and neighbors) to be used less frequently than in more familiar hazards such as tornadoes and floods (Drabek and Stephenson 1971). Instead, information would be sought from public safety personnel and others with expertise.

Consistent with these propositions, over half of the respondents at each site reported a frequent search--obtaining information about the threat four or more times a day. Less than ten percent of the sample reported getting volcano related information as infrequently as once a day. There is also some support for the proposition about information sources. Mass media (newspapers, television and radio) were mentioned by between 80-100 percent of the respondents at each site, while informal sources (friends, relatives and neighbors) were mentioned by about 70 percent of the sample. The only source of information that differed by site was contacts with emergency officials. These were reported by 65 percent of the Cougar residents, but only a minority of those in Woodland (17 percent) or Longview (9 percent).

Chronic threat: risk perception

By 1983, all residents were aware of the potential for continuing volcanic activity. Consequently, the chronic threat study assessed the salience of the threat in terms of the degree of disruption of household routines. Moderate or significant household disruption was reported by 48 percent of the Toutle residents, but only 19 percent of those in Lexington. The study also assessed the nature of the perceived community impacts of each of the two principal threats, ashfall and mudflow/flood. Respondents were asked to name up to three consequences for each of these threats. Respondents in the two communities were quite similar in their ability to name at least one consequence for each threat (Lexington: 97 percent for ash and 100 percent for flood; Toutle: 100 percent for both threats). Toutle residents were more likely than Lexington residents to report three consequences for each of the two types of threats (75 percent versus 58 percent for ashfall and 40 percent versus 24 percent for mudflow/flood). Residents of the two towns did not differ in the frequency of mention of health and safety consequences (ashfall: 48 percent in Toutle versus 46 percent in Lexington; flood: 35 percent in Toutle versus 31 percent in Lexington). The Toutle residents were more likely than Lexington residents to mention ash damage to buildings (96 percent versus 69 percent) and roads (41 percent versus 22 percent).

In judging the severity of the personal consequences of an eruption, respondents considered the consequences for property to be more severe than those for personal safety, a finding quite similar to that reported previously by Murton and Shimabukuro (1974), and Hodge, et al. (1979). There were, however, no appreciable differences between sites in terms of the residents' judgements of severe consequences for either personal safety (9 percent for Lexington versus 16 percent for Toutle) or property (44 percent for Lexington versus 47 percent for Toutle). In both communities, threat to property was significantly correlated with proximity to the river.

The chronic threat study also examined the extent to which families had developed evacuation plans that identified routes of egress and a safe destination. By the summer of 1983, 84 percent of the residents of Toutle had developed a family emergency plan. This is an increase from the 59 percent who had such a plan prior to the May 18, 1980 eruption (Perry and Greene 1983), and is significantly higher than the 56 percent of Lexington residents having a specific evacuation plan.

Chronic threat: information seeking

Residents of the two communities differed significantly in their responses to a question about their usual channels of information about the volcano threat. Toutle residents relied heavily on the radio (42 percent), somewhat less on newspapers (27 percent), and used personal conversations (18 percent) more than television (11 percent). Lexington residents tended to rely only on the mass media--newspapers (42 percent), television (36 percent), and radio (17 percent)--to the virtual exclusion of personal conversations (3 percent). The differences between the communities are even more noticeable in the responses to a question about the frequency of contact with each of a number of specific sources. For Toutle residents, the county Sheriff (93 percent), the county Department of Emergency Services (55 percent) and friends and neighbors (51 percent) were the only sources with a significant tendency to be contacted "sometimes" or "often." Lexington residents contacted a broader array of sources: the county Department of Emergency Services (81 percent), the county sheriff (79 percent), friends and neighbors (36 percent), the U.S. Geological Survey (35 percent), Army Corps of Engineers (32 percent) were the major sources described as having been contacted "sometimes" or "often."

Another question asked about sources of information regarding protective measures. In contrast to the patterns of information seeking about the volcano hazard, newspapers were rated as one of the principal sources of information about protective measures (29 percent in Toutle and 31 percent in Lexington). Personal judgement (32 percent in Toutle and 16 percent in Lexington) and emergency authorities (18 percent in Toutle and 32 percent in Lexington) were also reported to be useful sources. Other mass media and informal sources were rarely used for this purpose.

Conclusions

There have been considerable changes in the risk perceptions of residents in areas close to Mt. St. Helens. Those at risk are aware that the volcano is active, mudflow/flood and ashfall are the principal threats to their communities, and these threats can have a number of impacts that jeopardize personal safety and property. The volcano hazard has become salient, in the sense that local residents report frequent thought and significant disruption from it, and is now considered a serious threat--especially by those who live close to the volcano (ash threat) or to the river (mudflow/flood). The salience and severity of the threats, in turn, appear to have motivated the development of family evacuation plans, as well as the adoption of other adjustments (Perry and Lindell 1990b).

These studies of Mt. St. Helens document a complex pattern of information seeking that varies with the prior knowledge and experience of those at risk, proximity to the volcano, perceived expertise of information sources, and the characteristics of the channels of information transmission. They also suggest that risk area residents learned to differentiate among the information channels on the basis of the content of the information and the characteristics of the sources from which that information is received. In particular, these studies provide a basis for making a distinction among four types of hazard information: 1) cause and effect relationships in the behavior of the hazard agent (e.g., the knowledge that "If swarms of microtremors are observed, an eruption is likely to follow."); 2) current status of the threat (e.g., the knowledge that "swarms of microtremors are occurring now." 3) potential adjustments to the hazard (e.g., insurance, evacuation, floodproofing); and 4) recommended protective responses to the current threat.

The data from these studies suggest that local officials will be successful in enhancing levels of hazard adjustment to the extent that they can inform residents about the specific impacts to which they are vulnerable; emphasize the severity of the damage that could result from unprotected exposure and explain how potential damage could be reduced or avoided by the adoption of adjustments. Their efforts are particularly likely to be successful if this information is made sufficiently salient that people think about the hazard frequently. Of course, even frequent exposure to hazard relevant information does not automatically elicit attention and comprehension, let alone the acceptance, personalization and retention required to initiate hazard adjustments (cf. McGuire 1968; Mileti and Sorensen 1987; and Nigg 1982).

The finding that residents' perceptions of threat can increase even when they are unable to recall specific ways in which they could be affected seems to indicate that detailed hazard knowledge is not a necessary condition for initiating hazard adjustments. Consistent with this proposition, Lindell and Perry's (1992) literature review concluded that residents only need to believe that the threat really exists and protection is needed in order to motivate them to search for protective actions. Consequently, environmental cues, behavior of peers, and perceptions of information sources (especially the sources' expertise, trustworthiness, and attractiveness) could substitute for hazard related message content in promoting a perception of major threat. If overall threat perceptions were acquired directly from credible social sources, hazard related information would be needed only to understand the content of subsequent threat messages or to provide resistance to persuasion from counter-argumentation by dissenting authorities, experts or peers.

The Mt. St. Helens studies also suggest that examining citizens' perceptions of information sources and hazard adjustments is just as important as investigating their perceptions of hazard agents. Lindell and Perry (1992) and Perry and Lindell (1990b) have recently reiterated the observation that adoption of feasible adjustments may be substantially inhibited by local residents' lack of awareness of these adjustments; erroneous beliefs about their efficacy, cost, time requirements and barriers to implementation; and unresolved concerns about the available information sources. Further understanding the role of these factors is likely to make a significant addition to our understanding of the hazard adjustment process.

REFERENCES

Journal articles

Crandell, D. R., Mullineaux, D. R. and R. M. (1975) "Mt. St. Helens Volcano: Recent and Future Behavior." Science 187:438-441.

Drabek, T. E. and Stephenson, J. "When Disaster Strikes." Journal of Applied Social Psychology, 1:187-203.

Greene, M. R., Perry, R. W. and Lindell, M. K. (1981) "The March 1980 Eruptions of Mt. St. Helens: Citizen Information and Threat Perception." Disasters 5:49-66.

Mileti, D. S. (1980) "Human Adjustment to the Risk of Environmental Extremes." Sociology and Social Research, 64:327-347.

Perry, R. W. and Lindell, M. K. (1990b) "Predicting Long Term Adjustment to Volcano Hazard." International Journal of Mass Emergencies and Disasters, 8:117-136.

Perry, R. W., Lindell, M. K. and Green, M. R. (1982) "Threat Perception and Public Response to Volcano Hazard." Journal of Social Psychology, 116, 119-204.

Book references

Burton, I., Kates, R. and White, G. F. (1978) The Environment As Hazard, London: Oxford University Press.

Drabek, T, E. (1986) Human System Responses to Disaster, New York: Springer-Verlag.

Hodge, D., Sharp, V. and Marts, M. (1979) "Contemporary Response to Volcanism: Case Studies from the Cascades and Hawaii." Volcanic Activity and Human Ecology, Payton D. Sheets and D. K. Grayson, (eds.), New York: Academic Press.

Janis, I. L. and Mann, L. (1977) Decision Making: A Psychological Analysis of Conflict, Choice and Commitment. New York: Free Press.

Kunreuther, H., Ginsberg, R., Miller, L., Sagi, P., Slovic, P., Borkan, B. and Katz, N. (1978) Disaster Insurance Protection: Public Policy Lessons. New York: Wiley.

Lindell, M. K. and Perry, R. W. (1992) Behavioral Foundations of Community Emergency Planning. Washington, DC: Hemisphere Press.

McGuire, W. J. (1968) "The Nature of Attitudes and Attitude Change." The Handbook of Social Psychology, Second Edition. Gordon Lindsey and Elliot Aronson (eds.), Reading, MA: Addison-Wesley.

Mileti, D. S. and Sorensen, H. (1987) "Natural Hazards and Precautionary Behavior." Taking Care: Why People Take Precautions, Neal D. Weinstein (ed), New York: Cambridge University Press.

Murton, B. J. and Shimabukuro, S. (1974) "Human Adjustment to Volcano Hazard in Puna District, Hawaii." Natural Hazards: Local, National and Global, Gilbert F. White (ed.), New York: Oxford University Press.

Nigg, J. (1982) "Awareness and Behavior." Perspectives on Increasing Hazard Awareness, Thomas Saarinen (ed.), Boulder, CO: Institute of Behavioral Science, University of Colorado.

Perry, R. W. and Green, M. R. (1983) Citizen Response to Volcanic Eruptions. New York: Irvington Publishers.

Perry, R. W. and Lindell, M. K. (1990a) Living With Mt. St. Helens: Human Adjustment to Volcano Hazards, Pullman, WA: Washington State University Press.

Slovic, P., Fischoff, B. and Lichtenstein, S. (1980) "Facts and Fears: Understanding Perceived Risk.: Societal Risk Assessment: How Safe is Safe Enough, C. Schwig and William A. Albers (eds.), New York: Plenum Press.

Slovic, P., Kunreuther, H. and White, G. F. (1974) "Decision Processes, Rationality, and Adjustment to Natural Hazards." Natural Hazards: Local, National, Global, Gilbert F. White (ed.), New York: Oxford University Press.

Technical report

Earle, T. C., Southwick, L. L. and Lindell, M. K. (1981) Newspaper Coverage of Mt. St. Helens: Patterns of Content and Information Sources. Seattle, WA: Pattelle Human Affairs Research Centers, BHARC 411/81/035.

Unpublished paper

Lindell, M. K. and Perry, R. W. (1989) "The Protective Action Decision Model: Application to Technological Risk Perception and Communication." Paper presented at the workshop, Communication About Energy and Environment, Paris, France.

UNCERTAINTIES IN THE DEVELOPMENT OF PREDICTIONS OF EARTHQUAKES

V. KARNIK
Geophysical Institute, Academy of Sciences
14131 - Prague
Czech Republic

INTRODUCTION

First it must be clarified that "prediction of earthquakes" means usually the estimate of time of origin, place and magnitude of an event, whereas "prediction of hazard" means assessment of the intensity of impact (ground motion) caused either by a single extreme event or by the whole family of events affecting a certain locality. The second category includes "prediction of seismicity", i.e. the estimate of the level of future earthquake activity, e.g. in terms of the number of events of different magnitude and of the upper threshold magnitude event, during a given time interval.

Every category has its procedure containing steps of different accuracy, thus, the final product has only a limited validity and should be classified by a statement on uncertainty. Every attempt for prediction is always a broad interdisciplinary endeavour[1].

Prediction of earthquakes

Only the first case qualifies strictly as "earthquake prediction". There is a comprehensive literature on methodology of earthquake predictions, the methods applied are numerous, however, even a combination of all of them has not yet guaranteed a full success in predicting the time of origin, location of epicentre and magnitude of a potentially destructive earthquake. Successful predictions in China, the USSR or elsewhere, are very few as compared to more frequent failures. The reasons are in the inaccessibility of the medium in which earthquakes originate and in the great complexity of the earthquake generating process.The successful predictions that have been made so far have all been based on the observation and interpretation of "precursors", that is to say geophysical, geodetic, geochemical or biophysical phenomena, which are related to the earthquake

[1] N.B. The meaning of basic terms like hazard, risk etc. is different e.g. in seismology and hydrology, some standardisation of definitions is highly desirable.

J. Nemec et al. (eds.), Prediction and Perception of Natural Hazards, 167–172.

generation process and which have been observed to precede some earthquakes already in the past. Such precursors include e.g. abnormal seismic activity, ground deformation and tilt, water level changes and discharge from wells, anomalies in chemical contents of fluids in the earth's crust, geomagnetic and geoelectric anomalies, and abnormal behaviour of animals.

However, no precursor has yet been discovered which is an infallible indicator of the future occurrence of an earthquake. Furthermore, it appears that different precursors may appear at different stages in the earthquake generation process, i.e. there are short or long term precursors ranging from several minutes to several years, or precursory phenomena are detected in different localities in the earthquake preparation region, some localities seem to be more sensitive than others for reasons until now not well known.

In the absence of any one-to-one relationship between the occurrence of such precursory phenomena and the occurrence of earthquakes, it is desirable to adopt a certain "probabilistic" approach to prediction, based on the statistical analysis of seismic and other precursors. However, the volume of observational data has so far proved to be insufficient for valid conclusions to be drawn from such analysis, and all predictions, including the successful ones, have in general been based on collective judgement of the significance of the observed phenomena. Thus, the present state-of-the-art does not permit a general quantitative evaluation of accuracy of earthquake predictions because the whole problem is still at an early experimental stage and predictions can be evaluated only case by case considering intensity and number of precursors.

It is evident that predictions of various kinds will continue to be made throughout the world and that public authorities in countries subject to earthquakes will be faced with the problem of evaluating such predictions and deciding what action to take in response to them. It is therefore desirable that in each country where destructive earthquakes are liable to occur, the government sets up an Earthquake Prediction Evaluation Committee.There are two different viewpoints from which to consider response to predictions, namely that of authorities and institutions, and that of the public. There is a need to take any prediction (scientific or non-scientific) seriously as soon as it begins to attract significant public attention, or seems likely to do so. National or local government officials should watch out for such predictions and arrange promptly for an authoritative scientific review. Any preparedness planning which follows should be matched to the capacity of the majority of the population to respond to such plans.

All warning statements should include an explanation of the basis for issuing the warning as well as a statement on the reliability of the prediction. Official warnings should include specific recommendations for action by individuals or families. These actions must be appropriate for, and within the capability of, the majority of the population.

Government and scientific authorities should meet with media representatives at an early stage in order to establish procedures for accurate reporting of the situation as it develops.

It is also clear that without widespread public education about earthquake predictions, social disruption can ensue.

Governments should have public education programmes explaining in lay term what a prediction and warning means, how a prediction will be validated, how the public will be informed of its development, and what appropriate response they can take. Each country must naturally select the proper procedures corresponding to the political system and social and cultural background of the population.

Publication of predictions has on occasion given rise to undesirable consequences for communities in the areas of prediction.Scientists engaged in earthquake prediction activities should observe basic rules of behaviour, i.e. they should always inform research groups and national prediction committees of a country concerned before publishing any prediction.Warnings based on the evaluation must be issued only by government bodies, because only these bodies can take over the responsibility for all consequences of the warnings.Naturally, warnings can be issued only for regions surveyed by a complex observation systems monitoring various types of precursory phenomena. Establishment of such a system is quite expensive and observations must be carried out for years even decades to guarantee at least some degree of reliability of prediction. During the last three decades prediction networks became operational in test areas in the USA, the USSR, China, Japan, Turkey.

Seismic hazard and risk assessment

The estimation of future earthquake effects can be included also in the category of earthquake prediction because its integral part is the determination of future earthquake activity assumed to be valid for time intervals ranging from 25 years to 10.000 years.
Hazard and risk assessments represent significant inputs in earthquake engineering, land use planning and disaster management (preparedness) all aimed at substantial reduction of human and material losses.

The standard methodology of seismic hazard assessment has stabilized at present in the form which involves three basic steps:

a) definition of earthquake source regions,

b) determination of attenuation functions for strong ground motion including the site specific effects ("microzoning"),

c) probabilistic algorithm of hazard calculation.

Every step is composed of several operations and depends first of all on the availability of homogeneous earthquake information from the largest possible time span and on additional geological and geophysical data. Seismic hazard should consider the damage potential from two phenomena, i.e. from elastic ground motion as well as from ground failure and fault rupture. The sources of uncertainties are mainly:

a) poor inputs or the lack of some observational data on earthquake generation and seismic energy propagation,

b) the inadequate models of earthquake source (boundaries, N(M), (Mmax), of distribution of events in time (the models over-simplify the reality) and of local effects on attenuation of ground motion,

c) personal judgement of experts, diversity of experts opinion.

For consequences see the following table:

SOURCE	CONSEQUENCE
Short observation period covered by the catalogue	Extreme events may be missed, boundaries of source regions wrongly delineated, N(M) not representative, may cover either an active or a quite period.
Different magnitudes without unification	Large scatter in N(M), N(M) questionable
Use of m_b (short-period PV)	Saturation at M=6.5, bending of N(m), loss of distinction between large events, N(m) not usable for hazard assessment, errors in hazard values within ±1° or more.
Mean radii of isoseismals in site evaluation (instead of zimuthal attenuation).	
Different expert delineation of boundaries of source zones.	Errors up to ±1° in hazard.
Error in M_{max}	1/2 magn. unit results in 100% change of hazard values (the largest effect).
Doubling activity rate	Acceleration increases by about a third.

The faith in numbers produced by hazard assessments is great among users who have no sufficient knowledge of all the uncertainties involved and it is the duty of researchers to complete the numbers with an evaluation of the errors and, eventually, the diversity of an expert opinion. It is evident that under the present state-of-the-art, hazard assessments provide only orientational numbers which can guide the users, but which cannot be relied upon as exact values.

If hazard information is combined with information on vulnerability of various elements (buildings, lifelines, economy) exposed to hazard future losses (risk) can be estimated for a given area.

The uncertainties in loss estimates derive from several sources.First is uncertainty in the ground-motion intensity and ground failures for a given event. Second is uncertainty in estimating damage given the intensity and ground failures. Third is uncertainty in estimating the losses given damage to the facility. Finally, there is uncertainty in the process of inventorying the number of facilities in each building classification and geographic area.Each of these elements could be made more precise with additional effort and resources, but uncertainties are inevitable in any practical study of earthquake losses and should be expressed and quantified.

Transfer of knowledge and response of users

The mechanisms and rules for transferring individual earthquake prediction and response of authorities and population were discussed above. The process of earthquake loss mitigation requires the involvement of many specialist ranging from seismologists and geologists to development planners, economists, emergency planners and administrators. Without their co-operation no progress can be reached. All experts concerned must be aware of methods and data the others are utilizing and of all uncertainties involved. The transfer of knowledge from one professional group to the other is essential and efficient communication between them must be established and preserved. There are still barriers in mutual understanding caused e.g. by differences in terminology which is far from being standardized in the field of disaster management.

REFERENCES

UNDRO (1983): Proceedings of the Seminar on Earthquake Prediction Case Histories, Geneva 12-15 Oct. 1982.

UNESCO (1984): Proceedings of the International Symposium on Earthquake Prediction, Terra Publ. Co. Tokyo.

V. Karnik (1983): Earthquake prediction management. Paper presented at the Int. Earthquake Conf. Los Angeles 7-11 Febr. 1983.

4. .V. Karnik, I.L. Nersesov (1987): Principles and problems of assessment of seismic hazard. Proceed. Seminar on Earthquake Prediction and Mitigation of Earthquake Losses, Dushanbe, USSR, 8-14 Oct. 1986. UNDRO, Geneva.

USGS (1986): Proceedings, Workshop on "Probabilistic Hazard Assessments", Open-File Report 86-185, 1986, all papers, particularly by D.L. Bernreuther et al., p. 314.

NUREG/CR-1582, Vol. 3 (1980), Seismic Hazard Analysis - Solicitation of Expert Opinion, Washington, Aug. 1980.

FEMA (1989): Estimating losses from future earthquakes. Earthq. Hazard Reduction Series N° 51.

SESSION 5

THE USE OF SCIENTIFIC INFORMATION BY THE MEDIA

THE DIFFERENT WORLDS OF SCIENCE AND MASS COMMUNICATION: IMPLICATIONS FOR INFORMATION FLOW FROM THE FORMER TO THE LATTER

E. L. QUARANTELLI[1]
Disaster Research Center
University of Delaware
Newark, DE 19716
U.S.A.

This paper presents a perspective on a problem. While research observations are used to support my argument, there is no systematic setting forth of findings or summaries of studies on the topic. Instead, I mostly indicate the implications of my view.

The research base from which my remarks are drawn has recently been growing (for other than sources indicated later see Nelkin, 1987). Most of it deals with the physical sciences and engineering and primarily refers to the American scene. But I think my remarks are generally applicable in Europe and elsewhere where the social sciences exist and there is a modern mass media system.

My thesis is that the world of mass communication and the world of science are two different social worlds. The members of these worlds have different values and beliefs. This makes interaction between journalists and scientists difficult. The result is poor information flow between them, especially from the latter to the former. There is no easy solution to the problem but there are possibilities for improvement if those in science change some of their behavior in communicating with those in journalism.

The two worlds involved

Nearly three decades ago, C. P. Snow (1965) proposed looking at the natural sciences and the humanities as two separate cultures. The sociological specialty of the sociology of knowledge had even earlier proposed considering different areas of intellectual and work activities as distinctive social worlds (Merton, 1949). Such areas have their own particular set of norms, values, beliefs and expectations with respect to their central phenomena, whatever it may be. More recently it has been suggested that a sociology of knowledge approach should be interested not only with the social basis of intellectual productions, but

[1] Research Professor Disaster Research Center, University of Delaware, Newark, Delaware, 19716 USA.

J. Nemec et al. (eds.), Prediction and Perception of Natural Hazards, 175–182.

also the social consequences of knowledge. As such the conditions under which knowledge is produced, diffused and applied should be examined. Thus, there is:

> need for understanding the multiple frames of reference with which actors perceive knowledge and the discrepancies between the frames of reference of knowledge producers and knowledge users (Weiss and Bucuvalas, 1980: 302).

In this kind of framework, it is possible to talk of a scientific community and a comparable mass communication social institution, and the factors that will influence the kind of information flow that can occur between them. Our discussion indicates the nature of the two worlds and shows mostly the difficulty of transmitting scientific knowledge via the mass communication system.

Mass communication is the social institution or system of producing and distributing general information that uses a particular technology, the mass media. We distinguish, as increasingly is the case in the literature, between mass communication (MC) that has reference to the social aspects, and mass media that refers to the underlying technology. In more everyday terms, we have reference to the social organization involved in producing and distributing newspapers, television and radio programs, magazines, films, and the content of wire services and cable systems. The general information is intended for a non-specialized audience of fairly large size. Studies indicate that scientific information consists only a minuscule amount of what is transmitted daily by the MC systems, with radio coverage being the most extensive (Cronholm and Sandell, 1981: 86). But this does not include news reporting of disasters where at the height of such occasions in the United States and Canada, radio and television stations that operate may provide in their programs most of the information on the disaster (Wenger and Quarantelli, 1989).

With respect to disasters, MC systems can: 1) report news about proximate disaster happenings; 2) provide information about disaster planning; 3) undertake education about disaster behavior; and 4) project imagery about disasters.

As to the world of science, since it is presumably more familiar to readers of this paper, I will only say the following. The label of science here encompasses the generation and application of knowledge primarily obtained through scientific procedures. As such, the social and the physical sciences are included as are engineers and other technologically oriented practitioners.

A comparison

1. The MC area emphasizes reporting of the unusual or the different. Relevant here is the old story that it is not news when "dog bites man" but only when "man bites dog." The later is what is considered newsworthy. In contrast, scientists look for regularities or universals in whatever they study. The anomaly or the negative case may catch

attention but is then used to look for a different set of regular patterns. The norms operative in the two worlds on what is to be examined and communicated about differ, with emphasis on opposite ends of a continuum of unique-general.

2. Scientists try to root whatever they report in what others have found earlier. There is the whole scholarly apparatus of references, footnotes, bibliographies, review of the literature, etc., to inform users of the link between the present and the past. The MC world operates with a much more ahistorical framework. Thus, as Weiss has written:

> "In the media, each story is discrete, unique, isolated. Even if the same paper ran a story on directly related research a few weeks or months earlier, the new story almost never mentions it. The earlier research is not new; therefore, it is not news. There is little effort to go back and compare or synthesize... Media practice focuses on new... Newsroom procedures push toward timeliness, impact, significance, and uniqueness. Each report is a "story" in itself (Weiss, 1985: 40-41)".

3. There are differences in the two worlds as to who can handle information. The science world is one of specialists, of experts who are very knowledgeable but along very narrow lines. They do not see themselves as competent scientists outside their particular specialized spheres. Journalists do not claim any expertise but instead take the position that as generalists they can deal with what is needed. Thus, those in MC do not see stories about disasters as different from any other kinds of stories. They present no special problems, responsibilities or opportunities for the journalist (see Wenger and Quarantelli, 1989).

4. Both worlds are in one sense audience-driven, but in different ways as to expectations. Those in the MC area do not expect viewers or readers to provide any direct feedback. In fact, among American journalists there is a strong belief that they personally, as professionals, rather than their audiences can be the only judges of their work. Scientists are very attuned to what their peers will think and expect to be directly evaluated on the quality of their work. However, important for our purposes is not only that there are different audiences or reference groups, but different criteria of what is valid and worthwhile are used.

5. Conceptions of objectivity differ in the two worlds. It has been written that:

> "...mass journalism today implies a norm of objectivity that differs widely from the social scientist's understanding of that norm (Phillips, 1977: 73)".

The importance of this is not that there are two references of the term, but that different norms are involved. MC emphasis is on fairness; there is usually an assumption that there is a second or opposite side. Science stresses correctness, accuracy, validity, and not balance, impartiality or neutrality. As Sandman notes:

"Journalism, like science, attempts to be objective, but the two fields define the term very differently. For science, objectivity is tentativeness and adherence to evidence in the search for truth. For journalism... objectivity is balance. In the epistemology of journalism, there is no truth (or at least ways to determine truth); there are only conflicting claims, to be covered as fairly as possible (Sandman, 1986: 6)".

6. What is used to evaluate information is different in the two worlds. Practitioners in the MC area judge information in terms of their personal experiences and their understanding of the phenomena. The conception of having a "nose for news" rests on this ground. Scientists instead evaluate in terms of how the information has been produced, such as whether the right research techniques were used, and whether the results make sense in terms of existing knowledge. To some extent there is an "understanding" versus "knowledge" difference between the two worlds.

7. What is valued also differs in the two worlds. Timeliness is very important in the MC area, and thus a focus on the immediate. In contrast, in science the significance of the information is much more valued. Related to this is that scientific information is never seen as final or conclusive. Thus, scientists charge journalists write in ways that imply certainty.

8. The MC area attempts in various ways to personalize a story. Human beings are woven anecdotally into news stories in some way. In science, depersonalization is the norm as shown in stylistic conventions on writing style for professional journals. In fact, the traditional use of the passive voice in scientific accounts sometimes leaves unclear how anything happened!

9. The two worlds also differ in how they report information. The MC area strives for simplicity. Much of what is known is deleted. Scientific information tends to be complex with the use of jargon as an attempt to attain precision. Also, there is the use of many qualifiers. Given this, it is not surprising that judged from the scientific side, press reports on scientific information is sometime seen as incomplete or inaccurate.

10. There is a difference in the two worlds about the emphasis given to the abstract and the concrete. The MC area tends to use a logic of the concrete, setting forth specific "facts", and not venturing too far from them by staying at a descriptive level. It is almost as if "facts" speak for themselves", a position most scientists find very difficult to accept. Science, while it may deal with details uses them as part of pre-determined categories or building new categories. In this respect, the criticism of a news reporter of scientists she dealt with unwittingly had an element of validity to it; she said they "produce more theories than facts" (quoted in Glasser and Ettema, 1989: 14).

Implications

A major implication of our observations is that it will be very difficult to institute quickly any significant changes in MC reporting about any aspect of disasters. The reason is simple: the reporting process is a reflection of the very social structure of mass media organizations and of the subculture of the world of journalism (as is also true of everyday reporting, see Tuchman, 1978). The reporters, editors and other staff members involved in news gathering and other information flow have been socialized into a world and occupational subculture whose values, beliefs and norms are different from those in science. Journalists are not operating outside of some social setting. Just the opposite, they are following the framework of their world they have learned as a result of becoming members of MC organizations.

Social structures and subcultures of course are not static, and changes can be brought about in both. But alterations in the disaster relevant reporting of journalists in the field today will not be significantly altered by just telling them disaster mitigation measures are important, or that they should follow different field procedures in their coverage of disasters. Journalists will only consistently act differently if they are socialized or resocialized into different behavior patterns.

This in turn requires structural alterations in MC organizations and in journalistic subcultures. For example, changes in disaster reporting could follow, if as a result of preimpact planning, there was a strengthening rather than weakening of organizational gatekeeping during the emergency time period (actually newspapers, in contrast to radio and television, do increase their gatekeeping procedures, see Wenger and Quarantelli, 1989). Similarly, there might be alterations in emergency time reporting if, in the curricula of journalism schools, some negative consequences of disaster myths were systematically taught. One example is the situation where officials fail to issue warnings about threats because of a mistaken concern about generating "panic". Likewise, changes in reporting could result if journalistic norms and values stressing speed of reporting and "beating the competition" were downplayed for those emphasizing accuracy and gaining a reputation as a legitimate source--in more popular terms, a New York Times rather than National Inquirer approach. These are merely examples of possible ways of trying to influence the structures and cultures within which journalists operate (for other examples, see Wenger, 1985).

It is of interest that to the extent changes are advocated, it is frequently assumed that mass communicators should primarily change. But I am not too sanguine that prime reliance should be placed on such a strategy. There is not that much by way of incentives or payoffs for the MC area to give high priority to initiating major changes to improve information flow for disaster purposes. At a more complex level, MC organizations operate much like other organizations, which means their actions are characterized (if we take the work of the Noble Prize winner, Herbert Simon, 1976 seriously), by bounded rationality and satisficing. That is, groups limit even the relevant information they process and do not seek to maximize achievements of their goals.

Actually, it might be easier to institute changes on the scientific side. This is because there are operative social factors that could facilitate, if not encourage, changes in the information flow from science to the MC area, provided advantages was taken of them.

First, there are some success stories where we can see that scientific information is passed on relatively well by the MC system. In the United States the system does a respectable job in issuing warnings about certain kinds of disasters such as hurricanes and tornadoes. The situation is far from perfect, but most of the existing difficulties exist because of gaps in the communication flow from monitoring systems of scientists to mass media personnel. This stems from a structural problem of going from a state or governmental system over into the private sector; most of the problems here do not result from the above mentioned differences in the two worlds. But leaving the problematical aspects aside, the success partly stems from the actions of social scientists convincing National Weather Service personnel they were issuing warning messages, which while technically correct, were not meaningful for the general populace (see McLuckie, 1973; 1974).

Second, leaving aside any theoretical importance, local disaster planners need to get certain information out to various sectors of the public and different organizations. They have a strong motivation to do so. The better ones have good links with their local MC systems (see Drabek, 1985). What is being suggested is that there are social entities already in place that can serve as catalysts for change. Perhaps scientists ought to try to move such officials to try to influence the MC system in addition to interacting directly with it. They might make good surrogates.

Third, within the MC area there are some activities that come close to science or at least have a strong affinity with it. For example, there is in the United States the prestigeful occupation of investigative reporter (see Glasser and Ettema, 1989); there has also been the development of the specialization of science writers. The scientific area should see occupants of these roles as target audiences, a more specific focus than the general populace or MC generally. Investigative reporting is especially interested in most anything of a socially dubious nature or where there might be much social conflict. Certainly there are some aspects of disaster planning, whether in mitigation, emergency preparedness, response or recovery which scientists could call to the attention of investigative reporters. Science reporters are especially attuned to issues in their fields of specialization.

Fourth, disasters are inherently interesting to human beings, and most people will also agree that it is an important policy topic. Disasters are not esoteric matters of interest to only a few. This is an advantage this subject matter has over many others. Now some scientific information about disasters may not easily fall into such a category. But it is should be possible to link almost any information to the more interesting aspects of disasters. This will be particularly possible if it is remembered that the ultimate goal or use of the scientific enterprise is to save lives and otherwise reduce the tolls and disruptions of disasters; it is not just to increase scientific knowledge per se or to satisfy only the intellectual curiosity of scientists.

REFERENCES

Journal articles

Cronholm, M. and Sandell, R. (1981) "Scientific information: A review of research." Journal of Communication 31: 85-96.

Drabek, T. (1985) "Managing the emergency response." Public Administration Review 45: 85-92.

Glasser, T. and Ettema, J. (1989) "Investigative journalism and the moral order." Critical Studies in Mass Communication 1: 229-259.

Merton, R. (1949) "The role of applied social science in the formation of policy." Philosophy of Science 16: 161-181.

Phillips, E. (1977) "Approaches to objectivity: journalistic vs. social science perspectives." Pp. 346-357 in J. Yinger and S. Cutler (eds.) Major Social Issues. Beverly Hills, CA.: Sage.

Weiss, C. (1985) "Media report card for social science." Society 22: 39-47.

Weiss, C. and Bucuvalas, M. (1980) "Truth tests and utility tests: Decision-makers frames of reference for social science research." American Sociological Review 45: 302-313.

Book references

Nelkin, D. (1987) Selling Science: How the Press Covers Science and Technology. NY: Freeman.

Sandman, P. (1977) Explaining Environmental Risk: Some Notes on Environmental Risk Communication. Rutgers, NJ: Environmental Communication Research Program, Rutgers University.

Simon, H. (1976) Administrative Behavior. NY: Free Press.

Snow, C. P. (1965) The Two Cultures and the Scientific Revolution. Cambridge, MA: MIT Press.

Tuchman, G. (1978) Making News: A Study in the Construction of Reality. NY: Free Press.

Wenger, D. and Quarantelli, E.L. (1989) Local Mass Media Operations, Problems and Products in Disasters. Newark, DE: Disaster Research Center, University of Delaware.

Unpublished papers

Wenger, D. (1985) Mass media and disasters.

Technical reports

McLuckie, B. (1973) The Warning System--A Social Science Perspective. Fort Worth, TX: National Weather Service.

McLuckie, B. (1973) Warning--A Call to Action. Fort Worth, TX: National Weather Service.

EFFECTIVE COMMUNICATION BETWEEN THE SCIENTIFIC COMMUNITY AND THE MEDIA

BRUNA DE MARCHI
Commission of the European Communities
Joint Research Centre, T.P. 650
Institute of Systems Engineering & Informatics
I-21020 Ispra (VA)
Italy

Disaster and system vulnerability

The sociology of disasters has always attempted to reach a shared definition of its subject matter. The numerous critical reviews of the literature (e.g., Quarantelli and Dynes, 1977; Kreps, 1984; Drabek, 1986) identify and discuss a number of conceptualizations that are often quite diverse and sometimes incompatible. This lack of consensus shows the vitality of the discipline, which is able to keep an interest in theoretical debate, even under growing pressure to provide crude empirical data and simplified practical advice.

By means of a diachronic analysis, Quarantelli (1982) outlined a progressive and generalized trend from physically based conceptualization of disaster to socially based ones. Socially based conceptualizations focus on human system responses to hazardous events rather than on the disaster agent itself. The sociological study of disasters focuses on patterns of individual and collective action and looks for similarities and differences, even independent of the origin of the calamity.

The most recent and innovative theoretical contributions link the concepts of disaster and system vulnerability. With this approach, cooperation becomes possible between social scientists from different disciplines who have been working independent of each other on similar topics. Not only is vulnerability conceptualized as "a system's capacity to absorb and recover from the occurrence of a hazardous event" (Timmerman, 1981, p. 21), but it is also thought of as the degree of control that is (or is not) exercised over the event itself (Di Sopra and Pelanda, 1984).

As a physical event, an earthquake cannot be prevented, but as a disaster it can. The vulnerability of human systems to earthquakes, volcanic eruptions, droughts, tornadoes, hurricanes, floods, landslides, etc., can be reduced by promoting appropriate urban

J. Nemec et al. (eds.), Prediction and Perception of Natural Hazards, 183–191.

planning, the careful management of resources and effective education for civil protection. Deeply internalized cultural strategies for coping with disaster are as vital for the survival of human systems as reinforced concrete is for the preservation of buildings. They provide both individuals and social groups with a set of routinized defensive behavior patterns and attitudes for coping with recurrent danger (Moore, 1963; Anderson, 1965).

What now has to be accomplished is the operationalization of the concept of vulnerability and the construction of valid and reliable indicators to link this theoretical development to effective actions. This will enable the proper assessment of natural hazards and the design of successful responses. As Liverman suggests, vulnerability may be mapped "in geographical space (where vulnerable people and places are located) and in social space (who in a place is vulnerable)" (1990, 32). Almost half a century of research has shown that disasters hit the poor and weak more severely and with longer lasting effects than the rich and strong, be they individuals or groups or nations.

The concept of vulnerability, furthermore, enables both natural and man-made disasters to be treated within a common theoretical framework. If vulnerability is a feature of the human system and can, therefore, be either increased or diminished by human action, there is no such thing as a natural disaster. All disasters are, to a certain extent, man-made.

The concept of vulnerability also helps to concentrate the researcher's attention on the reasons why disasters occur--on "those events which are relevant to the sociological study of the causes and preconditions of disaster" (Turner, 1978, 83). Thus, a conceptualization of disasters in terms of system vulnerability is most useful when the practical goal is the design of warning systems for the detection and prediction of potentially disastrous hazards, either natural or man-made.

Warning information: issuing instructions or communicating?

Discussion of the problem of how to prepare and implement warning systems is beyond the scope of this paper. That task entails detailed technical knowledge and the expertise of different disciplines. Natural hazards must be identified and classified and warning signals occurring in nature must be detected and recognized as inputs to technological systems. The focus of this paper is on the dissemination of warning information, as it is practiced in the industrialized world. This process of dissemination is made up of different and often overlapping and intertwined steps, all of which have to be carefully analyzed and acted upon. The questions to be asked include the following: What is the legitimate source for issuing a warning? What is the most credible source? To whom should the warning be addressed? What should be its content and structure? Should there be a single message or different ones according to diverse audiences? How many times and how frequently should the warning be repeated? Which channels and media are likely to be the most effective?

To answer these questions (and many more), social relationships as well as technological tools must be taken into account. Elsewhere, I have maintained that the often observed unsatisfactory response to warnings, which is extensively documented in the sociological literature (e.g., Nigg, 1987), derives from failure to consider the relational aspects built into any communication (De Marchi and Ungaro, 1987; De Marchi, 1988). The former are not just passive message receivers, like machines in the mathematical theory of information (Shannon and Weaver, 1949) but active message interpreters. They are not, moreover, isolated individuals, but rather social actors involved in extended communication networks.

There are plenty of accounts of warnings which, although received by the target population, have not been appropriately understood and/or acted upon. At times, people do not recognize a warning as such, or they understand it but do not believe it, or they understand the "what is happening" section of a warning, but not the "what to do" section. In all such cases, what needs to be explored is a social relational failure, not a technological one. Questions need to be asked about the internal congruency of the message, its correspondence with the receiver's needs and mental model, the credibility of its sources and the possibility of interference caused by different messages coming from different sources.

Although sociological research has, on the whole, contributed a great deal to the understanding of how communication works in an emergency, the study of warnings still has a technological bias. Most such studies are inspired by the mathematical theory of information and seem to ignore other traditions in communication research (e.g., Watzlawick, et al., 1967; Bateson, 1972) and the refined models of semiotics that are better suited to representing human communication (Eco et al., 1965; Eco, 1975). In other words, the underlying assumption of most studies of warning seems to be that all unsuccessful communication derives from the failure of the technological equipment to disseminate messages that accurately reproduce "objective reality." Were it possible to overcome such inadequacy, they seem to imply, the rational behavior of individuals and organizations would follow almost automatically.

This perspective needs total revision. Warning information is aimed at promoting proper public behavior in case of an emergency. This goal cannot be achieved merely by issuing instructions. Its attainment must be based on a long-term strategy where a stable communicative relationship is established between the public and those in charge of managing the hazard and the aftermath of a disaster. Such a strategy will result in people being better equipped for receiving warnings and for actively seeking and performing life preserving actions (De Marchi and Mc Dermott, 1990). People do not simply follow instructions; they make decisions. Good warning information should facilitate access to previously acquired knowledge, thus improving people's capacity for making decisions (De Marchi, 1990; De Marchi and Rota, 1990).

The media and the scientific community

Several types of social relationship are implied in the dissemination of warning information. Some of these, such as those between the mass media and the general public, have been thoroughly explored. Others, such as those between the scientific community and the mass media, have been neglected. This failure is quite amazing, since the encounter between the scientists and the media actually generates the first input into the warning dissemination process. My hypothesis is that this failure results from the resistance by many scholars to seeing themselves (as members of the scientific community) as objects for observation and study.

It is, of course, more comfortable to blame the mass media for distorting a supposedly objective reality than to admit that such distortions may be the result of interaction between communicating social actors. To blame only the mass media, however, implies that the scientific community possesses a single, unquestioned and unquestionable truth that gets spoiled when it is communicated to the general public through the media. The latter, it is implied, purposefully distort the information to provide sensational news. If it were as simple as this, the solution would be equally simple (at least in some countries): a law, or equivalent legal measure, requiring the media to acknowledge official scientific statements as such and to quote or broadcast them with no cuts or changes.

As every scientist knows, things are not so simple. Science works with hypotheses that are often disproved and with theories that may be provisional. Scientists dealing with risk and safety issues have to cope with strong uncertainty and even ignorance (Funtowicz and Ravetz, 1990b) and are often called upon to express their judgements on "ambiguous facts" (Otway and von Winterfeldt, 1992). Since Galileo, the scientist's moral commitment has been to dismiss any theory that is empirically refuted. Scientists must be ready to abandon any previously held idea and to start their work again and again (Levi Montalcini, 1987). This is, of course, even more so when it comes to detecting and predicting potentially disastrous natural hazards.

It is not my intention to minimize the distorting power of the media. This power is documented by many valuable studies dealing mainly with the post-impact phase of disasters (e.g., Scanlon, et al., 1985). In particular, one has in mind the "disaster myths," propagated by the media and resistant, as any myth, to empirical falsification (e.g., Quarantelli and Dynes, 1972; Rosengren et al., 1978). Here I want to distinguish between different sectors and phases of media operations. The media are complex organizations with multiple actors, units, roles, functions, tasks, aims and goals. They also develop their own culture, attitudes, values and beliefs. They may transmit information, but they may also collect materials to be published or broadcast or they may prepare such materials. Different units of the media are active in the different phases and take care of the different tasks.

There are differences between and within the media. "At best, the media are not only essential communicating channels, but also critical watchdogs and sometimes virtual scientists themselves...At worst they are quick and dirty sensationalists" (Macgill, 1987, 47). In television, for example, there are commercial and noncommercial stations and specialized and nonspecialized ones; there is network and cable television; there are programs differing in content, structure and style, including the results of high quality investigative journalism. Maisel writes that "the radio networks, mass circulation magazines and general format radio died over a decade ago. TV, with the lead foot of cable TV on the throttle, is rapidly undergoing the same process" (1973). Although he is talking of the USA, the same trend is observable in many industrialized countries.

There is no such thing on "television" or "the media" as such. The first thing to take into account when warning information is circulated through the mass media is the nature of the audience that can be reached. If warning information is disseminated only through the news media, those members of the population who are not exposed to them will be automatically cut out from reception. The warning message must also be tailored to the intended receiver. In this phase it is not yet the media audience, but a media person who is collecting the information.

Looking now at the other participant in the communicative relationship that is the subject of this paper, namely the scientific community, some remarks seem appropriate. The first concerns the very word "community," since this leads one to think of a more homogeneous group than is actually the case. It may be assumed that scientists share a common general goal (the advancement of knowledge) and, to a certain extent, a common method. In practice, science is made up of different disciplines and each discipline includes diverse and sometimes conflicting schools. Scientists also perform different roles. They may, for example, be employed in official institutions or they may be academics or they may work as advocates for a recognized interest group. All this may result in differences of opinion when it comes to assessing a natural hazard. Scientists must always cope with strong uncertainty. Tentative hypotheses and probability assessments are an essential part of the scientific enterprise. Acknowledgement of the uncertainty built into certain judgements underlines the professional trustworthiness of the scientific expert. There is always the risk, however, that such an acknowledgement will be misunderstood and taken for lack of competence by a lay audience.

Scientists, as well as the representatives of the media, have their own professional agendas, including a number of personal and collective values and criteria. It is important for them to seek new funding for research, to popularize scientific knowledge and to disseminate life preserving information. At the same time, it is also important for them to preserve their professional integrity, their status as experts, a positive image of their discipline and a good relationship with the public. This is all but irrelevant when it comes to disseminating warning information.

Some strategies for effective communication

The scientists who communicate warning information to the media must first recognize their interlocutors' background, commitments, values, needs and expectations. Dealing with an investigative journalist, for example, requires a different style and different skills from those required when dealing with a reporter. Secondly, scientists must be able to recognize and define the context and setting of the interaction. Are the scientists being interviewed? Are they delivering an official statement? Are they participating in a scientific discussion, a political argument or a public debate? Are they involved in a news program or one devoted to scientific popularization or entertainment? Thirdly, the scientists must acknowledge and meet their interlocutors' agendas. Thus, if the media representatives are looking for novelty, this need has to be fulfilled. If the scientists do not include some element of novelty in their information, the media are likely to distort the information according to their need. Finally, the scientists must be attentive to the feedback they receive to check whether the message originally produced is the same as the one received and whether the definition of the communication setting is agreed upon.

All such steps, of course, require a number of skills, ranging from the selection of a common language to the capacity of understanding (in the sense of *verstehen*) the interlocutors or, more simply stated, of stepping into their shoes. The ability to simplify complex information without making it trivial is particularly important (Sandam, 1988; Funtowicz and Ravetz, 1990a). Pragmatically, it is not so relevant to ask whether the dissemination of accurate scientific information is a goal for the media. It may or may not be. The scientists' pragmatic aim should be that the available scientific knowledge is received by the target audience. Their efforts should, therefore, be directed at conveying information in a way that is useful to the target population and, at the same time, that appeals to the media. If the achievement of such a goal requires some adjustment to the traditional style of scientific communication, it seems reasonable to argue that "the end justifies the means."

The communication strategies outlined above for the scientists also apply to the media and the same skills are necessary. The representatives of the media seeking hazard information must first recognize their interlocutors' professional background, commitments, values, needs and expectations. Next, they must unambiguously define the character and context of the interaction. Thirdly, the media must acknowledge and meet the scientists' agenda. Finally, they must be attentive to feedback to check whether the definition of the communication setting is agreed upon and whether the needs and expectations of both parties are met. Again, all these steps require the development of a number of skills that can and should be refined.

One essential step for coping with hazards and risks is to reduce the vulnerability of human systems. Since appropriate information is essential for disaster prevention and reduction, improving communication skills within a social system means reducing its vulnerability and making it more robust and capable of coping with calamity. A priority task is to ensure

that the mass media are well equipped for effective communication with the public. To achieve this, social actors must start by communicating effectively among themselves.

REFERENCES

Journal articles

De Marchi, B., and Mc Dermott, I. (1990) "Communication Pour Temps de Crise." Preventique 35, 31-35.

De Marchi, B. (1988) "Comunicazione ed emergenze di massa: spunti da un'esplorazione della letteratura." Programma Emergenze di Massa, Quaderno 88-2, ISIG, Gorizia.

Kreps, G. A. (1984) "Sociological Inquiry and Disaster Research." Annual Review of Sociology, 10, 309-330.

Otway, H. H. and von Winterfeldt, D. (1992) "Expert Judgement in Risk Analysis and Management: Process, Content and Pitfalls", Risk Analysis, 12, 83-.93

Maisel, R. (1973) "The Decline of the Mass Media." Public Opinion Quarterly, 55, 2.

Quarantelli, E. L. and Dynes, R. R. (1972) "When Disaster Strikes (It Isn't Much Like You've Heard and Read about." Psychology Today 5, (February), 66-70.

Quarantelli, E. L. and Dynes, R. R. (1977) "Response to Social Crises and Disasters." Annual Review of Sociology 3, 23-49.

Sandam, P. (1988) "Explaining Risk to Non-Experts." Civil Protection 9 (Winter), 14-15.

Scanlon, J., Alldred, S., Farrell, A. and Prawzick, A. (1985) "Coping With the Media in Disasters: Some Predictable Problems." Public Administration Review 45, 123-133.

Book references

Anderson, W. A. (1965) Some Observations on a Disaster Subculture: The Organizational Response of Cincinnati, Ohio to the 1964 Flood. Disaster Research Center (DRC), Columbus, OH.

Bateson, G. (1972) Steps to an Ecology of Mind. Ballentine Books, New York.

De Marchi, B. (1990) "Assessing People's Information Needs About Major Accident Hazards: Improving Knowledge for a Better Response." Gow, H.B.F. and Otway, H.H. (eds.) Communicating With the Public About Major Accident Hazards. Elsevier Applied Science, London and New York, P. 389-400.

De Marchi, B. and Ungaro, D. (1987) "A Sociosystemic Model of Information Management in Mass Emergencies." Dynes, R.R., De Marchi, B. and Pelanda, C. (eds.), 119-133.

Di Sopra, L. and Pelanda, C. (1984) Eds., Teoria Della Vulnerabilità. Franco Angeli, Milano.

Drabek, T. E. (1986) Human System Response to Disaster. An Inventory of Sociological Findings. Springer Verlag, New York.

Dynes, R. R., De Marchi, B. and Pelanda, C. (1987) Sociology of Disasters. Contribution of Sociology to Disaster Research. Franco Angeli, Milano.

Eco, U. (1975) Trattato di Semiotica Generale. Bompiani, Milano.

Funtowicz, S. O. and Ravetz, J. R. (1990a) "The Communication of Quality of Hazard Information." Gow, H.B.F., Otway, H.H. (eds.) Communicating With the Public About Major Accident Hazards. Elsevier Applied Science, London and New York, 367-78.

Funtowicz, S. O. and Ravetz, J. R. (1990b) Uncertainty and Quality in Science for Policy. Kluwer Academic Publishers, Dordrecht.

Levi Montalcini, R. (1987) Elogio Dell'Imperfezione. Garzanti, Milano.

Macgill, S. M. (1987) The Politics of Anxiety. Pion, London.

Moore, W. E. (1963) Social Change. Prentice-Hall, Englewood Cliffs, NJ.

Nigg, J. M. (1987) "Communication and Behavior. Organizational and Individual Response to Warnings." Dynes, R.R., De Marchi, B. and Pelanda, C. (eds.), 103-17.

Quarantelli, E. L. (1982) "What is a Disaster? An Agent Specific or an All Disaster Spectrum Approach to Socio-Behavioral Aspects of Earthquakes." Jones, B.G. and Tomazevic, M. (eds.) Social and Economic Aspects of Earthquakes. Institute of Earthquake Engineering and Engineering Seismology, University of Kiril and Metodij, Skopje, YU, 453-78.

Rosengren, K. E., Arvidsson, P. and Sturesson, D. (1978) "The Barseback 'Panic'. A Case of Media Deviance." Winick, C. (ed.) Deviance and the Mass Media, Sage, Beverly Hills, 131-49.

Shannon, C. E. and Weaver, W. (1949) The Mathematical Theory of Information. University of Illinois, Urbana, IL.

Turner, B. A. (1978) Man-Made Disasters. Wykeham, London.

Watzlawick, P., Helmick Beavin, J. and Jackson, D. D. (1967) Pragmatics of Human Communication. W.W. Norton, New York, NY.

Monograph reference

Timmerman, P. (1981) Vulnerability, Resilience and the Collapse of Society. Environmental Monograph No. 1, Institute for Environmental Studies, University of Toronto.

Technical reports

De Marchi, B. and Rota, E. (1990) Risk Information Needs of Communities Near Seveso Sites: A Pilot Study. Commission of the European Communities, Report EUR 12887 EN, Luxembourg.

Liverman, D. M. (1990) "Vulnerability to Global Environmental Change." Kasperson, et al., Understanding Global Environmental Change, The Earth Transformed (ET) Program, Worcester, MA, 27-44.

Unpublished papers

Eco, U., Fabbri, P., Giglioli, P. P. et al. (1965) "Prima Proposta Per un Modello di Ricerca Interdisciplinare Sul Rapporto Televisione/Pubblico." Instituo di Etnologia e Antropologia Culturale, Perugia (mimeo).

IDEAS FOR A GLOBAL MODEL OF COMMUNICATION AND IDNDR POTENTIAL

MARCO LOMBARDI
Sociological Institute
University of Cattolica
Largo A. Gemelli 1
20123 Milan
Italy

INTRODUCTION

This paper has two aims. The first is to identify the role of communication in emergencies. The second is to stress the problems and difficulties emerging from the recent debate on these themes.

Definitions of the term "Emergency"

Discussion is limited to two definitions of an emergency. The first, **frequency**, is based on a probabilistic universe. It characterizes the event along a continuum ranging from high probability to low, or from **normality** to "non-normality." Using this classification, an emergency is a rare event.

The second definition, referred to as **cognitive**, assumes that the survival of the species is supported by a predictable environment. Added to this assumption is the social-cultural system concept of **dominium of the environmental variability**. This concept, through cognitive and technological means, seeks to reduce the whole normality, including rare events. This reduction does not reduce the probability of the disaster; but does enormously reduce the damage. In this context, an emergency is an event that cannot be cognitively controlled by the social system (Cattarinussi & Pelanda, 1981).

The waves model

The waves model describes an emergency as a number of sharply rising demands. Each of these demands (needs) describes a quick increment curve (or wave) that becomes flat and approaches normality. Depending on the adaptive answer curve, a corresponding answer curve is produced by the system. When the two curves (waves) coincide, each need is met in a timely manner. The eventual delta indicated by differences among the areas subtended

J. Nemec et al. (eds.), Prediction and Perception of Natural Hazards, 193–198.

to the curves, (waves) represents the **vulnerability of the system** and identifies the number of demands without an answer (Dynes et. al., 1987).

This representative model of an emergency is applicable to all the needs asked by a system. The crisis is represented by successive needs-curves or waves. Thus, this model is identified as the waves model.

Information and emergencies

In general terms, information is explained as a particular form of energy that efficiently defies the **entropy** of the social or organizational system. A larger quantity or better quality of information increases the organizational level of the system and decreases the levels of confusion and improbability. It is now appropriate to introduce the concept of **meaning**. As a variable, meaning is endowed with a degree of improbability relative to the object of the information. This improbability is a function of the object of the information and as a function of the selective and oriented user of the information process - the receiver of the information. Information as an energy flow is characterized by its own **intensity** value and by an oriented vector that defines the **sense** of information. Using this perspective, it is necessary to understand the function of the characteristics of the flow source. This means it is necessary to understand its criterion that, among the many contrasting information inputs, has presided over the **strategic** choices. These choices include: transmission of information, the orientation and manipulative practices of the mass-media and the sources. The receiver also has a criterion that is adopted to formulate an opinion and contributes to the meaning given to the communication. This requires the identification of a new elaboration process that examines each fact and indicates ones' relational context (Suzuki, 1991).

The qualitative characteristic of the message's **significance** becomes important. The meaning attribution characteristic of the receiver is also the strategic key on which sources and news broadcasters orient their own positions. There are some characteristics that can define the receiver. In general, the emotional/cognitive needs addressed by the mass-media refer to:

a) the necessity of acquiring information and knowledge used in social interactions;

b) the need to reduce tension and conflicts in the subject and in the social reality;

c) the need to maintain its own balanced cognitive system.

In the specific context of an ongoing emergency, a third characteristic is the necessity of acquiring a surrounding reality map. This map must guarantee sufficient information to maintain the elements of a balanced cognitive system and to reduce the possible dissonance produced by the emergency.

Having made that long and necessary introduction, I will propose a frame for a Global Model of Communication for Emergencies.

A global model of communication for emergency

The Global Model of Communication for Emergencies is organized in three levels:

- **Level One:** Description and understanding of the process by which a receiver cognitively defines the event as an emergency;
- **Level Two:** Description and understanding of the demand for information by the system during the emergency;
- **Level Three:** Identification of the monitoring system of the emergency and crisis; that is, the informative system of the crisis.

Level one

This hypothetical model of mass emergency communications describes data input to the receiver coming through both direct and indirect channels. The direct channel is the experience. The indirect channel is the mass-media organized communication. The data is processed within the system according to its cultural contest, it unique situation, the incorporated model. The system produces information that becomes what we call the **reality** of the situation - the experience function; and the **image** of the event - the communication function (Lombardi, 1988).

The next step is the definition, from the system, of a **truth** level that is functional and orientative of the adaptive answers. These answers initiate the return to the normal end of the range as the dynamics of the system adjusts to the emergency. In this context, the concept of **truth** assumes relativistic connotations. Because each system creates its own levels of reality, the distance between **the reality** (the experienced fact) and **the image** (the broadcast fact). Lacking either the experience (or the memory of a similar experience) or the communication of the experience, **the truth** coincides with one of the two remaining terminologies, whichever is comparable.

Level two

Previously we defined the emergency in terms of **frequency** and **cognition**. The cognitive aspect of the emergencyis is the area of interest in our study of the event and its communication. Of particular interest in the observations information as a constitutive element of the emergency. In a description of the emergency, according to the **waves model**, both inquiries and demands behave in a similar manner. The movement of the demand beyond routine levels characterizes the emergency. Thus, the emergency continues until the communication channels do not produce information, the elaborated and full meaning of the term data. Emergencies are also characterized by a data surplus. The data surplus generates a level of complexity that the cognitive model cannot manage. Data appears as fragments of a larger whole. In the **waves** model, the answer curve describes an area larger than the demand curve. In both cases, the result is the same - lack of information. This is the case for the cognitive model because in the process of meaning attribution there is not enough material. In the case of the frequency model, the model itself is unable to manage the wave of information from the emergency process.

According to the **waves model**, there are two possible information crisis:

a) **lack of data**: the demand curve is wider than the answer curve. This is the classic description of emergency system vulnerability;
b) **surplus of data**: the answer curve is wider than the demand curve. The information is redundant, unclear and more complicated and complicates **cognitive metabolism**,

Information is different from other social demands that share the same process characteristic. For example, a surplus of beds in a hospital relative to the number of injured people shouldn't cause a crisis. Attempts to decrease the crisis condition must observe the fundamental principle that providing data is an insufficient response. What is needed is data that are adequate and congruous to the demand (Lombardi, 1991).

Level three

Perhaps the best approach is to create an effective monitoring system for emergencies and crisis. The system could include a network of multi-purpose and dedicated satellites capable of observing the earth or specific regions. There are such networks already operating.

<u>Communications - Planetary Networks</u>: Intelsat (15 satellites), Immarsat (6 satellites), and Intersputnik (3 satellites)

<u>Communication - Regional Networks</u>: Eutelsat (3 satellites, Europe, North Africa, Near East), Arabsat (2 satellites, Arabian System), Palapa (2 satellites, South East Asia) and Panamsat (1 satellite, Americas and Europe)

<u>Other Types of Networks</u>: Meteo Systems {Goes, Tiros and Noaa (USA), Cosmos (USSR), and GMS (Japan)}; Earth Sensing Systems {Cosmos, Spot and Landsat}; Position/Navigation Determining Systems (GPS (USA) Tsulada (USSR), Glomass, Geostar, Locstar.

Currently satellite acquired information is not very useful in reducing the global vulnerability of the system. There is potential for improvement of crisis management through the existent multi-purpose networks of satellites. In the future, new dedicated networks and land-based technological stations, flying stations and echo-satellites with remote sensors are feasible. Other developments might include:

- new protocols for the existent worldwide network of communication services;
- "high power" satellites and micro antennae that will cut the individual cost of joining the global network;
- broadcasting of emergencies due to technological developments and TV station interest;
- increased use of satellites for emergency communication;

Other projects include:

- Fly Away, a project involving the Red Cross and UNDRO to develop 40 small fly away Intelsat terminals;
- Star Project, the promotion of social and educational use of communication satellites (e.g. the Global University, Emergency Communication Services);
- Beamsat, a study by the Economic Commission on Europe-UN on new integrated technologies using satellites to monitor environmental risk. These satellites use multi-spectrum sensors, side looking radar, vidicam bean cameras, etc.;
- Cospas/Sarsat System, a project specifically for communication management in emergencies. Based on the USA and USSR networks of satellites, it is available to other nations. (Canada, Brazil, France, United Kingdom);
- Intelligent Beeper, a development of the Geostar System that consists of terminal stations that address early warnings and emergency instructions.

This is the general frame of the Global Model of Communications for Emergencies. It contains many problems; more problems than answers from a theoretical and technological point of view. However, I believe that today the scientific community has the opportunity offered by a new and great resource: the IDNDR program.

CONCLUSIONS

Inside the United Nations IDNDR main projects, scientists can cooperate at the global political level. This is the level that has the duty to manage emergency communication, to manage crisis and, first of all, to protect populations from the consequences of emergencies. From a general point of view, this also includes the responsibility to develop a new way of political life: we must all work towards a new model of integrated echo-politic. From a specific point of view, a sociological one, we know that the management of a crisis involves a normal process upset by the emergency. Intervention in the process has a strong impact on the dynamics of the affected social system. This intervention is a point of possible and deliberate interference with the evolving response process. In the critical stage of return to **normality**, (the return to the balanced conditions that the system defines as **normal**) the communication becomes an important factor in reducing or amplifying those dynamics. Most important in the global analysis of management dynamics problems is the understanding of the relationship between the media (information broadcaster and informal control institution) and the political actors and the decision system (information and formal control institution).

As I look around, I see three factors supporting the concept of a Global Model:

1) people are well disposed to use (people often ask to use) the communication and information system at a high level of complexity;

2) the technological response to this interest is a well integrated global communication system;

3) the reduction of vulnerability, civic and environmental, in itself is great media. Truly, in this case, "the media are the message."

There are two additional points I want to stress:

1) we must create a new political and institutional frame for the development of the Global Model of Communications for Emergencies;

2) we must produce protocols and procedures to link the three levels of the Global Model.

To me, these questions seem typical for a social scientist who has the duty to integrate all the levels of a new Global Communication system. To do otherwise is to create the possibility that the System will become only a wonderful technological toy.

References

Cattarinussi, R.R., Pelanda, C. (1981). "Disastro e Azione Umana", F. Angeli, Milano.

Dynes, R.R., Pelanda, C., De Marchi, B. (1987). "Sociology of Disasters", F. Angeli, Milano.

Lombardi, M. (1988). "Gestione dell'informazione nelle Emergenze di Massa", in: "Studi di Massa", Vita e Pensiero, n. 2, pp. 216-227, Milano

Lombardi, M. (1991). "L'appoccio sociologico alle emergenza", in: G. Bocca (cur.) "Prodigi paure ragione. Eventi Naturale Oggi", Guerini, Milano

Suzuki, H. (1991). "Study of Disaster Warning. Perspective", the Bulletin of the Institute of Journalism and Communication Studies, University of Tokyo, n. 44.

QUANTITY OF REPORTING ABOUT HAZARDS: THE CASE OF NATURALLY OCCURRING RADON

ALLAN MAZUR
Maxwell Graduate School of Citizenship and Public Affairs
Syracuse University
Syracuse, New York 13244
U.S.A.

People have intellectual and visceral motives for their behavior. The first is often weaker than the second, which is why many dieters succumb to the sight of food. Environmental hazards may be viscerally motivating too, when they are immediate and threatening, like the thunder and lightning of a storm, or a recent earthquake, to which people usually make reasonable alleviating responses (Quarantelli and Dynes, 1972). But most environmental hazards are too distant or improbable to be viscerally threatening -- for example, naturally occurring indoor radon, or future earthquakes. We know intellectually that these are potentially harmful, but they do not evoke anxiety. Therefore, though risk analysts tell us that as rational calculating actors we ought to protect ourselves, most of us do nothing.

When people do attend to a distant or improbable hazard it is usually because publicity in the mass media makes the hazard so salient that rational protection takes a higher than usual priority, or because publicity makes the hazard viscerally threatening, or both. This is nicely illustrated by radon, which has been known since 1979, if not earlier, to cause 5,000 to 20,000 lung cancer deaths per year in the United States (Mazur, 1987). Yet hardly anyone checked their household radon level until 1985, when the hazard suddenly became a frequent topic in the press. Remedial actions by governments, entrepreneurs, and a few homeowners quickly followed. By 1989, radon stories were far fewer in the mass media, and the various remedial actions lessened too (Sjoberg, 1989). Why did radon suddenly rise to prominence and then recede as quickly?

In December, 1984, an engineer named Stanley Watras, who worked at a nuclear power plant near Philadelphia, began setting off newly installed radiation alarms when he entered the plant. An investigation by plant technicians determined that the source of radiation was the extremely high concentration of radon in Watras's home. They notified Pennsylvania's Department of Environmental Resources (DER).

J. Nemec et al. (eds.), Prediction and Perception of Natural Hazards, 199–202.

By coincidence, the DER had recently received - and ignored - warnings from two technically-trained citizens that indoor radon was a serious problem in the area. DER personnel, astonished at the magnitude of radiation in the Watras home and now embarrassed that they had ignored the prior warnings, acted vigorously to test radon levels of other homes in the neighborhood. The Watras home quickly because a major story in the local news media, as did the possibility that nearby homes might have a similar problem. Neighbors were distressed that their families also might be in danger. Kay Jones, a homemaker, encouraged by a reporter, formed an organization called Pennsylvanians Against Radon.

As the story grew, the DER expanded its radon testing, checking over 2,000 houses by midsummer and finding that roughly half had unacceptably high levels. (This and later discoveries of unexpectedly high radon in certain locales have not much affected estimates of the total exposure of the American public to radon or to the total mortality.) These houses are located on the Reading Prong, a geological formation of granite, covering parts of three states. Rich in uranium, the Reading Prong was the presumed source of radon, suggesting that other homes on it also may have radon problems. By May, local press coverage diminished as testing became routine and the hazard picture stabilized. Perhaps the open windows of springtime also relieved some worries.

This alarm in Pennsylvania, the vigorous actions of the DER, and the reports by local journalists, all went unnoticed outside the area. No national-level news organ picked up the radon story for nearly four months after the Watras discovery. In contrast to the many local news organs operating in every city and town, there are a few major organizations, collectively called "the national news media," that have the resources and connections to cover national and international events regularly. The New York Times is preeminent among these, but also included are the major television networks, Time and Newsweek magazines, and the major wire services. Every day the national news media collectively produce a menu of stories from which thousands of local organs select their news of the day. Sometimes stories are simply repeated as they come in, and sometimes they are embellished. A story originating from a local news organ is unlikely to attain widespread coverage unless it is first picked up by the national media.

The New York Times picked up the Pennsylvania radon story from Robert Yuhnke, an environmental lobbyist who had worked for years on radiation issues. In early 1985, Kay Jones, the homemaker who formed Pennsylvanians Against Radon, contacted Yuhnke to ask his advice about her radon exposure. He then volunteered to talk to Philip Shabecoff, an environmental reporter for the Times, whom he worked with on other stories. Shabecoff knew nothing about radon prior to Yuhnke's call, but within a week went to Pennsylvania to report the story.

On May 19, 1985, five months after the Watras discovery and as the story was fading in Pennsylvania, the Times ran Shabecoff's report on the first page of the Sunday edition. It detailed Watras setting off alarms at the nuclear power plant and the subsequent discovery of radon in his home. Neighbor Kay Jones was reported to be "angry and frightened and frustrated."

Following Shabecoff's story, staff people at the United States Environmental Protection Agency leaked him a preliminary internal proposal to increase the Agency's work on indoor radon. This became the basis for a second story, on May 24, including the document's claim that radon in buildings causes 5,000 to 20,000 cancer deaths in the U.S. each year. I have already noted that such numbers date back to 1979 at least, but a reader could easily conclude from the Times article that this was a new risk estimate based on Pennsylvania's experience. As coverage continued in The New York Times, the basic story was carried almost intact by Time (July 22) and other magazines. By 1986 the story was spreading rapidly through all levels of the mass media, including television-network evening news (Mazur, 1987).

Radon was not newsworthy as a simple hazard. It became news when the risk was personalized in the human interest story of Stanley Watras, the engineer who set off radiation alarms as he entered the nuclear plant instead of when leaving it. There was also the pathos of Kay Jones and other mothers, fearful for their children. There is more fascination in stories of people than of toxins. The story was further personalized for people who lived on the Reading Prong, for they were identified as especially at risk, heightening their concern about radon.

Although journalists play a necessary role in calling any issue to public attention, it is simplistic to claim that the media alone placed radon on the public's agenda of risks. The sources who provide the "news" are equally important. Journalists cannot sustain a news story for very long if there is nothing new to report. Many local news organs tried to keep the story alive by sponsoring radon surveys in their areas, thus creating their own news. However, this quickly lost its novelty. Radon died as a story after 1988 simply because there was little new to report. This is why news organs seem to be continually identifying new hazards, which are briefly prominent and then fade away, making it difficult to sustain public attention on any one of them.

I regard the quantity of coverage about a hazard to be the primary vehicle whereby risk information is communicated to the public. With radon, when media attention was high, public response to the hazard--although never very high--certainly increased; when media coverage fell, alleviative actions dropped off as well. The same thing appears to occur with rises and falls of media attention to earthquakes (Turner, Nigg, and Paz, 1986) and to many other technological hazards (Mazur, 1990).

I emphasize the importance of quantity rather than the specific content of risk messages in the media. It is my view that substantive content of articles is not especially important in its effects on an audience. Consumers of news usually do not absorb articles in detail; instead they fashion simple images of hazards, often from headlines or pictures. Certain simple but vivid hazard images, such as California falling into the sea as the result of an earthquake, become especially widespread. I suspect that for many people, the simple image of radon - appearing so suddenly in the news - was of another "hazard of the month," which would soon drop from vogue (as, indeed, it did). Certainly the fear of radon affected few people viscerally, even at the height of publicity, and it is not surprising that relatively few people did much about it.

REFERENCES

Journal articles

Mazur, A. (1987) "Putting Radon on the Public's Risk Agenda." Science, Technology, & Human Values 12: 45-66.

Mazur, A. (1990) "Nuclear Power, Chemical Hazards, and the Quantity of Reporting." Minerva: 294-323.

Quarantelli, E. and Dynes, R. (1972) "When Disaster Strikes It Isn't Much Like What You've Heard and Read About." Psychology Today 5 (February): 67-70.

Sjoberg, L. (1989) Radon Risks: Attitudes, Perceptions and Actions. U.S. Environmental Protection Agency. Washington DC.

Book references

Turner, R., Nigg, J. and Paz, D. (1986) Waiting For Disaster. Berkeley CA: University of California Press.

EARTHQUAKE PREDICTION AND INFORMATION TO THE PUBLIC - A MEXICAN PERSPECTIVE

ROBERTO MELI
Director for research,
National Center for Disaster Prevention (CENAPRED)
Mexico City
Mexico

INTRODUCTION AND SCOPE

The aim of this contribution is to present some aspects of the relationship between the scientific community, the authorities, the press and the general public, regarding the prediction of possible disasters. The comments refer to the situation in Mexico and more specifically to two examples related to the seismic risk resulting from the high probability of a great earthquake in the so called "Guerrero gap".

Some peculiar characteristics of the situation in Mexico affect the earthquake prediction problem:

a) The population is highly sensitive to any news about the possibility of an earthquake The memory of the 1985 disaster is still vivid among most inhabitants. The frequent earthquakes of moderate intensity that are felt in Mexico City originate panic, especially among the dwellers of multistorey buildings.

b) The attitude of most communication media is characterized by sensationalism and often by ignorance about the technical and scientific matters they are dealing with. The information given to the media by the specialists is commonly misinterpreted and highly exaggerated.

c) The population generally distrusts the official information and is not eager to accept the measures proposed by the authority. The contradictory information often given by different official sources is partially the cause of this situation.

As a result of the aforementioned conditions the society is very sensitive to rumours and the problem of handling scientific information becomes very critical as can be appreciated by the two following examples.

J. Nemec et al. (eds.), Prediction and Perception of Natural Hazards, 203–212.

Earthquake prediction and the Guerrero Gap

Although some attention has been paid in Mexico to the development of methodologies for earthquake prediction, most seismologists agree that the possibilities of arriving to a short-term prediction are still very low. The instrumentation now in place in the area of major seismic activity and the careful observation of different phenomena allow to hope that, some time in the future, precursory activity could be identified for each of the major source mechanisms.

At present time, however, the only generally accepted type of prediction is in probabilistic terms, based in the theory of the seismic gaps. A regular pattern of major events characterizes most segments of the subduction zone in the Pacific Coast that is the source of the major seismic activity in the country. It can be assumed that the time between events and the magnitude of the earthquake corresponding to each segment, have a characteristic value, as can be seen from Table 1. Nevertheless the variability of both characteristic values is quite large, especially due to the uncertainty about the length of the fault that will break in a single event.

Region	Approximate Location		Year of Earthquake	Avg. Period
	Lat(*N)	Long(*W)		
E. Oaxaca	16.2	95.8	1897(7.4); 1928(7.7); 1965(7.8)	34
C. Oaxaca	16.0	96.8	1870 (7.9); 1928 (8.0); 1978(7.8)	54
W.Oaxaca	16.6	97.7	**1854(7.7); 1894(7.4);** 1928(7.4); 1968(7.4)	38
Ometepec	16.5	98.5	1950(7.3); 1982(6.9, 7.0)	32
San Marcos	16.7	99.2	**1845(7.9);** 1907(8.0); 1957(7.7)	56
Petatlàn	17.3	101.4	1943(7.7); 1979(7.6)	36
Michoacàn	18.1	102.5	1911(7.9); 1985(8.1)	74
Colima	18.4	103.2	1941(7.9); 1973(7.5)	32

Table.1: Observed Recurrence Periods of Large, Shallow,Interplate Earthquakes Along the Mexican Subduction Zone. The years in bold are less reliable (From Singh. et. al., 1981)

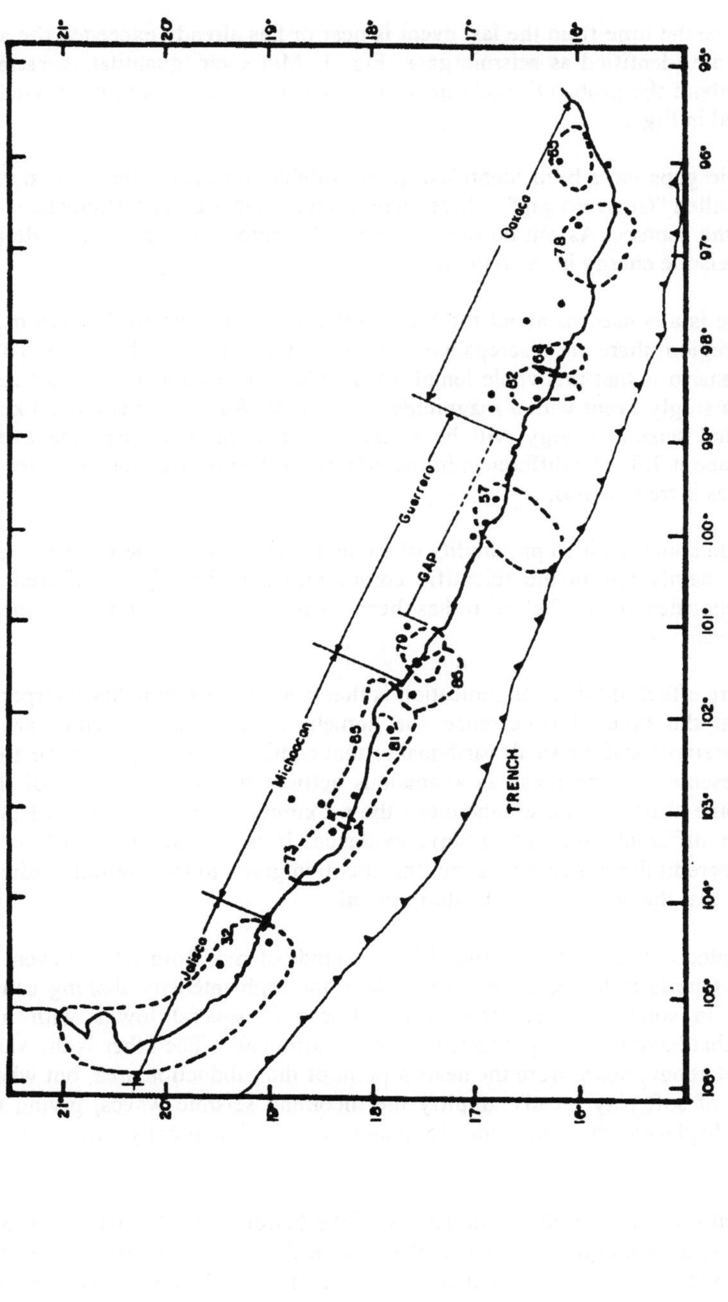

Fig. 1: Epicenters of major earthquakes in the subduction area of the Pacific Coast of Mexico between 1928 and 1986. Areas of rupture of some earthquakes are indicated by dotted lines(Adapted from Singh and Ordaz, 1990)

Segments where the time from the last event is near or has already exceeded the average return period are identified as seismic gaps, Fig. 1. Moreover, quantitative estimations can be made about the probability of rupture of each segment in some time window, as those presented in Fig. 2.

Several seismic gaps have been identified in the subduction area. The most significant one is the so-called "Guerrero gap", where there has not been a great earthquake since the beginning of the century. As can be seen from Fig. 3, there is a clear current deficiency of release of seismic energy in the region.

Although there is a consensus about the high likelihood of a great earthquake in a near future in this region, there are discrepancies about the magnitude of the event. The most worrisome scenario is that the whole length of the Guerrero segment, of about 230 km, will break in a singly event with a magnitude of about M=8.2. It seems more likely that the accumulated seismic energy will be released in two or three separate events of magnitude of about 7.8. The difference in the effects on the construction facilities by the two possibilities is tremendous.

The information about the high probability of a major earthquake in the Guerrero gap has been handled mainly within the scientific community and directly transferred to the concerned authorities. Nevertheless it has been also communicated to the media in several instances.

The most severe effect of this communication is that it has been frequently interpreted by the public that the time of occurrence can somehow be predicted with exactitude. Therefore the tremors and the small earthquakes that continuously happen in the area, are sometimes presented by the press as anomalous activity that is precursory of the big event. Also some charlatans have announced the proximity or the exact date of the "big one" based on different kinds of esoteric evidence. It is a continuous task of some scientists to discredit these predictions, but the attention given to the - initial prediction is always greater that the one given to the disapproval.

It must be pointed out that there are two different kinds of risk from a large event in the Guerrero gap. One is in the epicentral area where the high intensity shaking can cause great damage in weak rural constructions and also in several towns with modern constructions that have never experienced a severe earthquake. The other is the valley of Mexico sited at about 280km from the nearest point of the subduction area, but where the thick deposits of soft clay greatly amplify the incoming seismic waves, giving rise to large ground displacements with long dominant periods that greatly affect multistory buildings.

There is a great difference in the vulnerability of the buildings in the two areas. Mexico City has suffered a recent great earthquake that caused the collapse of nearly one hundred constructions and the significant damage of more than a thousand. This event has

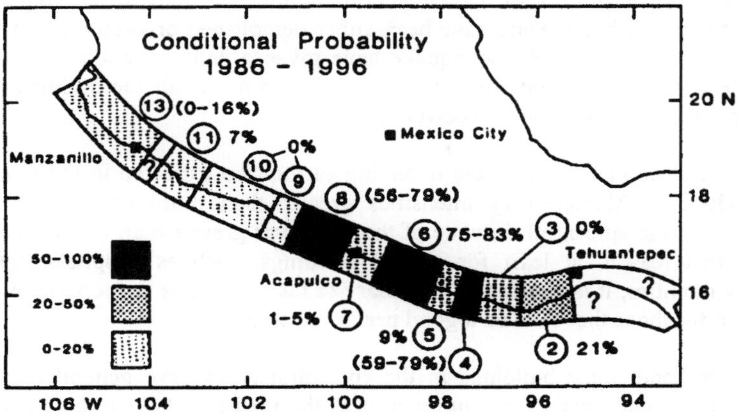

Fig. 2: Map View of the conditional probability estimates along the Mexican subduction zone for the time interval 1986-1996. Probabilities are based on our estimates of the expected recurrence time using either historically observed data or recurrence behaviour extrapolated from other segments of the margin. Encircled numbers refer to segments as discussed in text. Percentages in parenthesis represent values which, in our opinion, are less well-constrained. Segmentation of the margin into separate source zones based on the lateral extent of the last large or great earthquake in each segment. Question marks denote segments with insufficient data for analysis. (From Nisenko and Singh, 1987).

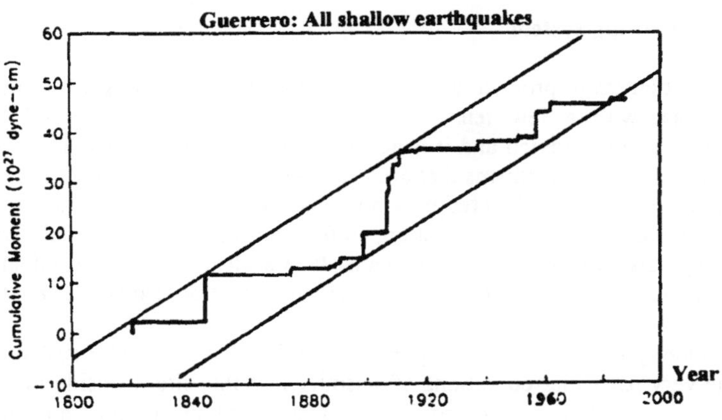

Fig. 3: Cumulative seismic moment release curve for Guerrero. Catalog of the past century's earthquakes appears reasonably complete. The slope of the parallel lines enveloping the data, 0.39×10^{27} dyn cm per year, suggests that seismic slip equals plate slip and gives a current moment deficit of about 20×10^{27} dyn cm. (From Anderson et. al., 1989)

constituted an important test for the whole building lot. The weakest buildings have been identified by the earthquake and have been either demolished or strengthened; those that were undamaged by the 1985 earthquake have given a demonstration of good seismic safety. The critical task in this region is to be able to detect all the buildings that have shown some sign of distress and to correct it.

Another critical point is to ascertain if the big earthquake expected in the Guerrero gap could produce in Mexico City intensities greater than those of 1985. It must be considered that the epicentral distance of the 1985 was greater than that to the Guerrero area. For this reason, at least for critical buildings, such as hospitals, schools and emergency facilities, it is considered necessary to take measures to increase their seismic safety, even for those that showed a good performance in 1985.

On the other hand, most buildings in the epicentral areas have generally experienced several moderate earthquakes, but never near to the intensity that can be reached by the one that is expected from the rupture of the gap. Few measures have been taken up to this moment to identify the most vulnerable structures and to retrofit at least the most critical ones.

For the aforementioned reasons the policy of most academic institutions and of many researchers has been to down play the prediction of the earthquake with the general public, in order to avoid false interpretations, and to stress to the authorities the need for a stricter control on the new constructions and for retrofitting vulnerable buildings.

The earthquake warning system

There have been attempts to produce devices that could give a warning of an incoming earthquake at least with a few tens of seconds of anticipation. One of them is commercially available in Mexico and it is actually sold in supermarkets. These devices are generally very simple mechanisms that are triggered when a threshold of acceleration is exceeded. In order not to be triggered by vibrations due to traffic, they must be set at a relatively high level of minimum acceleration, so that the chime will start only when the motion is so large that it can be felt by humans. It is a clear case of a product that is bought by the public - assuming that it will accomplish a function that is not able to do.

A serious warning system is now being prepared for Mexico City. The system is only suitable for the very particular conditions of seismicity of this site. Considering that the large intensity ground motions in Mexico City are due to earthquakes generated along the Pacific Coast at a distance of at least 280km, the seismic waves take a significant time to travel the distance. Therefore the ground motion can be detected in the epicentral area well in advance to when it starts in Mexico City. Assuming a typical velocity of the more harmful S waves of velocity, v, of 3.6 km/s, the time, t, they take to travel a distance, d, of 280km, is

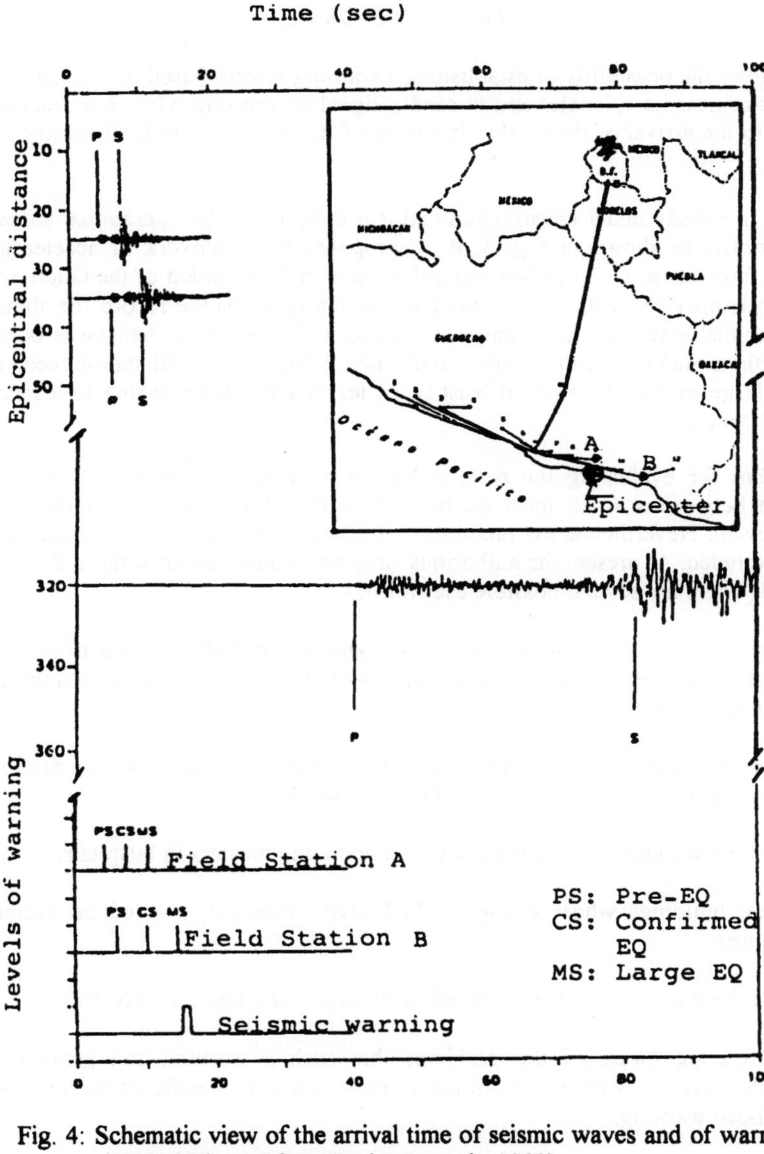

Fig. 4: Schematic view of the arrival time of seismic waves and of warning signals (Adapted from Espinoza et al., 1989)

$$t = d/v = 280/3.6 = 78 \text{ sec.}$$

This fact allows the possibility of establishing a warning system based on the detection of the earthquake near its epicenter and to send a signal of warning with about one minute of advance to the arrival of the motion in Mexico City. The concept is illustrated in Fig. 4.

The system is already under construction and it is expected to be operational sometimes in 1991. Briefly, as shown in Fig. 5, it is composed by a network of accelerographs deployed at intervals of about 25 km along the coast, in the segment of the Guerrero gap. Each station is provided with a microcomputer which analyzes the record for about five seconds after the arrival of the S waves, to evaluate if a major earthquake is occurring and to send the signal to a central station in the area. When the central station receives the signal of a large earthquake from at least two sites, it sends the warning to a receiving station in Mexico City.

The algorithm for evaluating the records has been carefully designed to be able to discriminate between a small local earthquakes with a high acceleration and a large magnitude event. Nevertheless the possibility of a false warning or of a missed warning cannot be excluded. At present the authorities have not decided about what to do with the alarm signal. There are several possible uses for it.

a) To send the warning to the subway system and to order all the train to stop at the next station in order to avoid their stopping within the tunnels due to an interruption of electric power.

b) To interrupt some critical industrial operations where the earthquake can produce an accident due to explosion or escape of toxic gases, for instance.

c) To send the warning to critical activities as operating theaters in hospitals.

d) To warn buildings whose occupants had been intensively trained on evacuation procedures.

e) To give the warning to the general public through radio and TV networks

The authorities are aware of the problems that such a warning can generate in a population that has not been trained to react properly and especially of the effects of a false or a missed warning.

The best policy seems to be to use the signal only to affect physical systems as in cases *a* to *c*, and in some experimental way in some well trained groups of cases *d* and *e*. The difficulty will be in limiting the access to the signal to the general public. The existence of the system cannot be a secret and there will be strong pressure from many

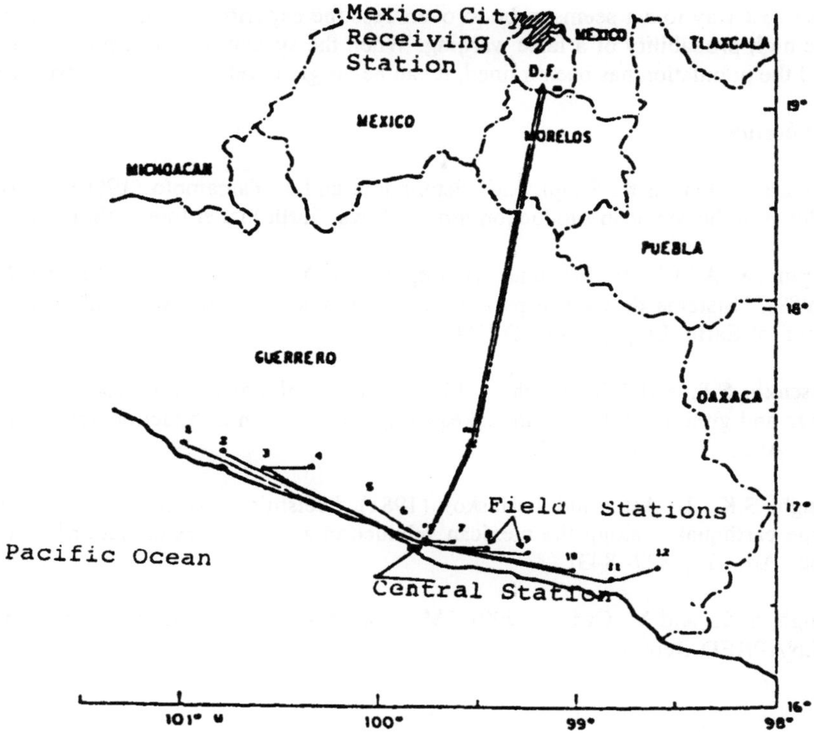

Fig. 5: Arrangement of the stations for the seismic warning system (Adapted from Espinosa et al., 1989)

organizations to have it available.

The best way to act seems to be to overstress the experimental nature of the system and the high probability of a false warning. When the system has been proven to be reliable and the population has been trained, it can be progressively opened to more users.

References

Anderson, J.G., S.K. Singh, J.M. Espmndola and J. Yamamoto (1989). "Seismic strain release in the Mexican subduction thrust". Phys. Earth Planet Inter., 58, p. 307-322.

Espinoza, A.M.J., B. Frontana, G. Legaria, S. Maldonado, A. Uribe and M. Medina (1989). "Sistema de control para el diseqo de una - alarma seismica". Proc. VIII Nat. Conf. of Earth. Eng., p D-98, D-lU4.

Nisenko, S.P. and S.K. Singh (1987). "Conditional probabilities for the recurrence of large and great interplate earthquakes along the Mexican Subduction zone". Bull. Seism. Soc. Am., 77. p. 2095- 2114.

Singh, S.K., L. Astiz and J. Haskov (1981). "Seismic gaps and recurrence periods of large earthquakes along the mexican subduction zone; a re-examination". Bull. Seismic Soc., Am., 1, p 827- 843.

Singh, S.K., and M. Ordaz (1990). "Movimientos fuertes en Mexico: una vision actual" CENAPRED Report.

Advances in Natural and Technological Hazards Research

Series Editor: Prof. Dr. Mohammed I. El-Sabh, *Département d'Océanographie, Université du Québec à Rimouski, 310 Allée des Ursulines, Rimouski, Québec, Canada G5L 3A1*

Publications

1. S. Tinti (ed.): *Tsunamis in the World.* Fifteenth International Tsunami Symposium (1991). 1993 ISBN 0-7923-2316-5
2. J. Nemec, J.M. Nigg and F. Siccardi (eds.): *Prediction and Perception of Natural Hazards.* Symposium Perugia, Italy (1990). 1993 ISBN 0-7923-2355-6

Kluwer Academic Publishers – Dordrecht / Boston / London

MIX
Papier aus verantwortungsvollen Quellen
Paper from responsible sources
FSC® C105338

If you have any concerns about our products,
you can contact us on
ProductSafety@springernature.com

In case Publisher is established outside the EU,
the EU authorized representative is:
Springer Nature Customer Service Center GmbH
Europaplatz 3, 69115 Heidelberg, Germany

Printed by Libri Plureos GmbH
in Hamburg, Germany